SOME ADAPTATIONS OF
Marsh-nesting Blackbirds

MONOGRAPHS IN POPULATION BIOLOGY

EDITED BY ROBERT M. MAY

Frontispiece. A male Yellow-headed Blackbird on his territory, Columbia National Wildlife Refuge, Washington, May 1977 (photo by E. N. Orians).

SOME ADAPTATIONS OF

Marsh-nesting
Blackbirds

GORDON H. ORIANS

PRINCETON, NEW JERSEY
PRINCETON UNIVERSITY PRESS

1980

Copyright © 1980 by Princeton University Press
Published by Princeton University Press, Princeton, New Jersey
In the United Kingdom: Princeton University Press, Guildford, Surrey
ALL RIGHTS RESERVED
Library of Congress Cataloging in Publication Data will be
found on the last printed page of this book
This book has been composed in VIP Baskerville
Clothbound editions of Princeton University Press books
are printed on acid-free paper, and binding materials are
chosen for strength and durability
Printed in the United States of America
by Princeton University Press, Princeton, New Jersey

Preface

An important thrust of ecological theory concerns itself with models whose predictions are derived from Darwin's theory of natural selection, the most general of all models of adaptation. The theory of natural selection in its general form is of limited usefulness, however, in explaining most specific cases. What is required is a series of models of intermediate generality which provides sharper foci into the mode of operation of natural selection upon specific traits. The way in which these models differ from those concerned with dynamics of populations has been clearly stated by Lack (1954a, pp. 4-6). In Lack's terminology models of adaptation concentrate upon the *ultimate factors* which have led to the evolution of attributes presently exhibited by organisms. In contrast, models of population dynamics take the attributes of individuals (or populations) as given and analyze their consequences for short-term fluctuations in response to environmental perturbations. Lack has termed this approach a *proximate factor* approach. Elsewhere (Orians, 1962) I have presented an argument for the importance of this distinction for understanding many of the controversies which have been so characteristic of the history of ecology.

It is clear that both approaches have contributed much to our understanding of ecological systems, and both continue to be employed for appropriate purposes. For the most part, adaptive models are of recent origin, and though theoreticians have dealt with these kinds of problems only a short time, testing of the models already lags behind. It is important that empirical verification or falsification of theories about adaptations be carried out in conjunction with their formulation if theoretical developments are not to deviate excessively from reality. In addition,

vii

since these ideas provide powerful bases for the development of predictive ecological theory, they are of immense practical as well as theoretical importance.

Models of natural selection are of two general sorts. One type predicts the behavior and/or morphology of individuals. These are essentially individual optimization models, and they make no assumptions other than the operation of classical Darwinian selection. Their concern is with which phenotypes should be most fit in certain kinds of environments. The second type may be termed population models in which statistical properties of populations are predicted. Some of these models follow directly from models of individual fitness or kin selection, but others require the operation of interdeme or group selection (Williams, 1966). The properties and problems of these models of selection have been reviewed by Hamilton (1964), Maynard Smith (1964) and Brown (1966). The real problem is to find methods by which the action of interdeme and group selection can be identified in the field. The easiest method would presumably be to deduce demonstrable consequences of interdeme or group selection. In practice, however, this is extremely difficult because unambiguous predictions from these theories are not easily made. Moreover, there are many traits which can be explained by group or interdeme selection but which are also open to explanations from Darwinian selection (Williams, 1966). In these cases the predilection of the observer normally dictates which explanation is favored.

In the present study I attempt to test models that predict four types of characteristics of individuals: habitat selection (Levins, 1968; Fretwell and Lucas, 1969), foraging behavior (Charnov, 1973, 1976a, b; Charnov, Orians, and Hyatt, 1976; Emlen, 1966; Krebs, Ryan, and Charnov, 1974; MacArthur and Pianka, 1966; Schoener, 1971); territoriality (Brown and Orians, 1970; Fretwell and Lucas, 1969;

Orians and Willson, 1964), and mate selection (Orians, 1969, 1972). The varied social systems of blackbirds and the structural simplicity of their foraging environment make some aspects of all of these models testable in the marshes of western North America.

The book is organized into an introductory chapter describing my study systems, a chapter describing the resource base supporting breeding blackbirds in western North America, two chapters dealing with adaptations of blackbirds to these resources (Habitat Selection, Foraging Behavior), two chapters on patterns resulting from these adaptations (Variability in Use of Resources; Competition, Overlap and Community Structure), a chapter presenting comparative data on Argentine marshes and blackbirds, and a final chapter examining the structure of marsh bird communities in a more general perspective.

Throughout I use theory in two distinct but complementary ways. One is the familiar practice of designing tests of theoretical constructs that allow rejection or provisional acceptance of their validity. The other is to use theory as a means of gaining insights about what should be examined in nature. It will become evident that this second use of theory has been more important in my studies because available theories concerning the ecological traits I examined are highly simplified and are not readily tested in the complex, uncontrolled field conditions in which I worked. Even if blackbirds are behaving in accordance with theory, the observable patterns will be modified by factors not included in the theory; results, therefore, are open to several interpretations. Nonetheless, during the project, theories repeatedly suggested measurements I would not otherwise have thought of, so that my overall understanding of adaptations of blackbirds has been substantially improved.

This study has benefited from the contributions of many people to both fieldwork and development of theory. In particular I wish to thank my students (Eleonora H. D'Arms, Robert K. Furrer, Celia R. Holm, Henry S. Horn, William Searcy, Mary F. Willson, and Douglas Wood), who used the opportunity provided by the blackbird project to contribute their own ideas to the ongoing theme as they were in turn enriched by it. Also vital was the help in the field of George Clarke, Paul P. Cook, John M. Emlen, Lynn Erckmann, John Falkenberry, Richard S. Fleming, Carlyn E. Orians, Kristin J. Orians, Dennis R. Paulson, Eric R. Pianka, John C. Schultz, James Simpson, Susan M. Smith, Lowell R. Spring, Perry R. Turner, and Jared Verner. From beginning to end, the aid of Lynn Erckmann with typing, reading proof and preparing illustrations, has been invaluable. The development of the concepts presented was aided, in addition to the many contributions of the above-mentioned people, by conversations with Eric Charnov, Lynn Erckmann, Henry S. Horn, Daniel H. Janzen, Richard Levins, Robert H. MacArthur, Robert May, Charles Monnett, Eileen O'Connor, Robert T. Paine, Nolan Pearson, Sievert Rohwer, Steven Syrjala, Michael Turelli, and John Vandermeer. Permission to work on the wildlife refuges and assistance with small problems that arose were provided by Earle Brooks, David Brown, Edward Collins, Philip Lehenbauer, and Jon Malcolm. From its inception, while I was a graduate student at the University of California, Berkeley, the entire project has been generously supported by funds from the National Science Foundation. Fieldwork in Argentina was made possible by a John Simon Guggenheim Memorial Fellowship.

Gordon H. Orians
Seattle, Washington
1978

Contents

SOME ADAPTATIONS OF
Marsh-nesting Blackbirds

CHAPTER ONE

The Approach
and the Subjects

1.1. THE PLAYERS

The blackbird family (Icteridae) is composed of about 96 species of birds breeding from central Alaska and northern Canada to the southern tip of South America. Though they are believed to be monophyletic and of fairly recent origin (Beecher, 1951), they nonetheless exhibit a range of social organization types which is paralleled in the avian world only by the varied social systems of the weaver birds (Ploceinae). Many of the icterids are medium-sized birds (30-100 grams) of open habitats and are often the most abundant birds in the areas in which they breed. This makes them especially well suited for investigations in which comparative data from a large number of individuals is necessary for testing theories. Finally some species, particularly the Red-winged Blackbird (*Agelaius phoeniceus*), are serious crop pests in parts of their ranges (Neff and Meanley, 1957). Therefore valuable data from extensive studies, not available for economically unimportant species, can be used to supplement results from my studies with a stronger theoretical orientation.

Among the icterids there are forest-, brushland-, grassland- and marsh-inhabiting species. Many of the species, including those I have studied, though primarily granivorous as adults, feed their young chiefly on insects. Spacing patterns among blackbirds range from large territories, within which all the resources of the breeding pair and their young are obtained, to dense colonies. Mating patterns range from monogamy, through varying degrees

3

of polygamy, to well-developed promiscuity. Finally, the group includes cowbirds which are brood parasites.

The present work concentrates on the two most abundant marsh-nesting blackbirds of western North America, the Red-winged Blackbird and the Yellow-headed Blackbird (*Xanthocephalus xanthocephalus*). The upland-breeding but marsh-associated Brewer's Blackbird (*Euphagus cyanocephalus*) was also studied simultaneously during the period of most intensive work on Redwings and Yellowheads in Washington (Horn, 1968, 1970; Furrer, 1974), and I supplemented those data with some of my own. In addition, I did a one-season comparative study of three Argentine marsh-nesting species, the Scarlet-headed Blackbird (*Amblyramphus holosericeus*), the Yellow-winged Blackbird (*Agelaius thilius*), and the Brown-and-yellow Marshbird (*Pseudoleistes virescens*).

The Red-winged Blackbird ranges as a breeding species from east-central Alaska and the Yukon south to northern Costa Rica, and from the Atlantic to the Pacific (Figure 1.1). It is migratory at the northern parts of its range, but many winter as far north as southern Canada. Over most of its range it is the most abundant breeding passerine bird in marshes, which are its usual nesting habitat. In parts of its range, particularly in eastern North America, it has, in recent years, spread rapidly as a breeding bird in upland pastures and crops.

The Redwing is strongly territorial, each male defending an area of marsh or upland within which up to one dozen, but usually no more than several, females construct their nests. Territoriality is only weakly developed among females. Individuals of both sexes feed extensively off their territories. Correlated with the polygynous breeding system (Selander, 1966), there is considerable sexual dimorphism in plumage and size, the males in eastern Washington weighing about 65-80 grams, the females about 35.

4

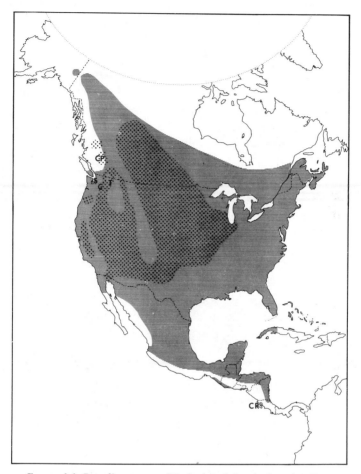

FIGURE 1.1. Breeding range of Red-winged (hatched) and Yellow-headed (stippled) Blackbirds. Within its range the Redwing breeds in nearly every marsh, and, in the East, upland habitats are extensively utilized. Only high-productivity lakes are occupied by Yellowheads throughout their breeding range. Locations of my main study areas are indicated by letters: S = Seattle, C = Columbia National Wildlife Refuge, T = Turnbull National Wildlife Refuge, CP = Cariboo Parklands, British Columbia, CR = Taboga, Costa Rica (adapted from Bent, 1958).

The males arrive on breeding areas first and establish territories prior to the arrival of females. Nest construction, incubation, and nearly all of the feeding of nestlings is done exclusively by the females. Rarely do male Redwings in Washington feed nestlings, but they commonly feed fledglings. During the nonbreeding season Redwings are highly gregarious, traveling in flocks and normally roosting colonially over water. Birds that breed in eastern Washington winter mostly in California but some remain in the Columbia Basin, chiefly around cattle-feed lots. Western Washington birds are resident.

The larger Yellowhead has many biological features in common with the Redwing. Males are strongly territorial, mating is polygynous, and nest construction and incubation are done entirely by the females. However, male Yellowheads, while participating less than females in feeding nestlings, do so much more regularly than male Redwings, usually feeding the nestlings at the first nest to hatch (Willson, 1966). The territory is a more important source of food than for Redwings. Males weigh about 95-105 grams and females about 60-65 grams.

The breeding range of the Yellowhead is restricted to the arid and semiarid areas of North America from south-central British Columbia, north-central Alberta and Saskatchewan and southern Manitoba south through the Great Plains and Great Basin to northern Arizona, central New Mexico, northern Texas and Oklahoma (Figure 1.1). Yellowheads are absent west of the Cascade Mountains in Washington but breed sparingly in the Willamette Valley of western Oregon. They appear again west of the Cascade-Sierra axis in the lowlands of California south to the Imperial and Lower Colorado Valleys. To the east the breeding range is roughly congruent with the former boundaries of the Prairie Peninsula in Wisconsin, Illinois,

Indiana, and Missouri, but sporadic breeding occurs farther east (Sawyer and Dyer, 1968). During the winter Yellowheads travel in flocks in central and southern California, southern Arizona, New Mexico, Texas, and the Mexican Plateau to its southern extremity. Recently a few have wintered in eastern Washington. The first males arrive in eastern Washington in mid-March, over a month after Redwings have become common. Further north, however, the difference in arrival times is less and male Yellowheads arrive in Saskatchewan only two weeks after Redwings (Miller, 1968).

The Brewer's Blackbird, by contrast, is nonterritorial, colonial and monogamous in eastern Washington. There is less sexual dimorphism in both size and plumage, males averaging about 70 grams and females about 60 grams. The species is widespread as a breeding bird in western North America from southern Canada to northwest Baja California, central Arizona, central New Mexico and northern Texas, and from the Pacific coast east to the western Great Lakes. In recent years its range has been expanding eastward, and there are now records of breeding as far east as southern Ontario (Devitt, 1964). During the winter Brewer's travel in flocks and range from extreme southern Canada (sparingly) south to central Mexico. The sexes arrive at the breeding sites already paired.

The Tricolored Blackbird (*Agelaius tricolor*), which is virtually restricted to lowland California as a breeding bird, is the most densely colonial of all North American passerines; colonies range up to 200,000 nests (Neff, 1937; Orians, 1961; Payne, 1969). In appearance they are almost indistinguishable from Redwings, and they are also highly gregarious during the nonbreeding season. Territories of the males are very small, averaging only a meter or two on a side, and all foraging for food for the young takes place

7

outside the colony area. Nest building and incubation are done exclusively by females, but males regularly feed both nestlings and fledglings.

1.2. THE THEATER

The present study has concentrated upon birds during the breeding season when, in my study areas, they are largely dependent upon marshes. Even though there is evidence that some bird populations may be influenced by the availability of winter resources (Errington, 1945; Lack, 1966; Fretwell, 1972) restriction of this study to the breeding season is appropriate for several reasons. First, the mating and spacing patterns of interest are breeding-season phenomena only. Second, in Redwings and Yellowheads there is a substantial floating population of males during the breeding season, indicating sufficient winter survival to saturate the breeding environment. There is no evidence of a floating population of females, but this is much more difficult to detect. Third, since starvation is a common form of nestling mortality, selection for foraging efficiency by the breeding adults should be very strong.

A temperate zone marsh is an unusual environment whose properties have had a strong influence upon the evolution of the eco-behavioral traits of blackbirds. First, the major source of food for insectivorous but nonaquatic birds in a marsh is emerging aquatic insects. These are present in great abundance for a limited time, but during much of the year a marsh provides little animal food to a bird unable to forage under water. In addition, this food supply is a rapidly renewing one. The availability of prey on day $x+1$ is almost uninfluenced by predation on day x because emerging insects harden and fly away from the water within a few hours. Consequently, the availability of food the next day is determined by factors, such as weather conditions and general productivity of the lake for aquatic insects, which affect the quantity of emergence that day. Even within a day, particularly during peak hours of emergence, the rate of renewal of insects is so high that a given spot can profitably be revisited many times in rapid succession by the same or different birds. This influences several strategies of blackbirds, including foraging behavior, habitat selection and mating patterns.

The capacity of marshes, from acid bogs on the one extreme to nutrient-rich lakes of semiarid areas on the other, to produce emerging aquatic insects, ranges over many orders of magnitude. Even within a small area, lakes may differ strikingly in their productivity as a result of differences in local drainage systems and depth. The most productive lakes and marshes are generally those in drier regions where they often have no outlets and therefore concentrate nutrients entering from the surrounding watershed. In addition, these lakes normally lack fish, which when present may depress the quantity of emergence of aquatic insects to very low levels. However, in many arid areas some lakes completely dry up during the summer, which eliminates them as habitats for many of the most important insect foods of breeding blackbirds.

Though blackbirds are primarily dependent upon marshes and the insects produced by them in my study areas, the nature of the adjacent uplands is also of importance. Recently emerged (teneral) aquatic insects leave the water as soon as they are able to fly but may be captured in adjacent upland vegetation. In addition, during periods of the day when few aquatic insects are emerging, or during bad weather, which inhibits emergence even on productive lakes, food available on uplands may be important for nestling survival.

Variations in marsh vegetation are also significant to birds. At one extreme salt concentrations may be sufficient

9

to prevent growth of emergent aquatic plants. At the other, in regions where the lakes have outlets and, hence, constant water levels, vegetation growth usually results in solid, dense stands of emergent vegetation 2 or more meters in height. In very humid areas, especially in cooler regions, a floating mat of vegetation often forms, effectively eliminating penetration of sunlight into the water. For this reason, mild to moderate disturbance, by breaking up dense stands of emergent plants, increases the number of emerging aquatic insects.

For all these reasons, marshes are environments which vary strikingly in space and time. Because recent models deal with adaptations to varying environments, marshes provide good conditions for testing such models. The availability of a wide variety of marshes in a small area with the same climate and weather has enabled me to study the impact of variability without introducing the complications which inevitably accompany studies along broader geographical gradients. Temporal and spatial patchiness of marsh environments vary in scale from diurnal to yearly and from microhabitat to regional changes. The general nature of this patchiness and the aspects of blackbird ecology they influence are summarized in Tables 1.1 and 1.2. In general, the larger the scale of the patchiness, the less often a decision is made by an individual with reference to it and the greater the probability that the response is a morphological one that commits the individual over its entire life. An important variable, weather, is not entered on Table 1.2 because it ranges from rough to fine grain and, hence, may affect all attributes of blackbirds. Its role will be explored more fully in subsequent chapters.

Fieldwork was carried out at four major locations in the Pacific Northwest, two of which are located in the region of large Miocene lava flows of the Columbia Basalt formations of eastern Washington (McKee, 1972). The Columbia Na-

TABLE 1.1. Spatial grain of the environment and attributes of blackbirds affected.

Type of grain	Scale	Blackbird attributes affected
Geographical	Regional differences in properties of marshes and uplands	Habitat preference General morphology Clinal variation in behavior
Coarse	Larger than individual foraging areas or territories	Territory size (males) Selection of mate and territory (females) Colony site location (Brewer's) Phenotypic variability of populations
Fine	Smaller than foraging areas of individual birds	Decisions about foraging mode, patch selection and prey choices Choice of nest site

tional Wildlife Refuge (The Potholes) is located in the heart of the Columbia Basin desert. Its topography consists of basins and buttes scoured in basalt by Pleistocene floods when the present channel of the Columbia River was blocked by ice (Bretz, 1959). It lies in the rain shadow of the Cascade Mountains and receives only about 200 mm of precipitation annually. Rainfall is generally light during

TABLE 1.2. Temporal grain of the environment and blackbird attributes affected.

Type of grain	Scale	Nature	Blackbird attributes affected
Rough	Yearly	Vegetative succession Changes in water levels Differences in quantity of emergence of insects	Phenotypic variance— niche breadths Territory location and size
Coarse	Seasonal	Emergence patterns of insects Vegetative growth Water levels Predator activity	Breeding season limits Clutch size Nest locations
Fine	Diurnal	Emergence patterns of insects Temperature Light	Foraging areas and choices of individual prey

11

the blackbird breeding season, but strong winds are common. The dominant upland vegetation of the area is sagebrush (*Artemisia tridentata*) and bunch grass (*Agropyron spicatum* and *Poa* spp.) (Figure 1.2), but, due to heavy grazing, native grasses have been largely replaced by the annual cheat-grass (*Bromus tectorum*). Rabbit brush (*Chrysothamnus nauseosus* and *C. viscidifolius*) is common on sandier soils and Hopsage (*Grayia spinosa*) grows on very dry ridges. Greasewood (*Sarcobatus vermiculatus*) dominates on alkaline flats together with Saltgrass (*Dactylis glomerata*).

Basins in the area were mostly dry prior to the initiation of agriculture at the turn of the century. Extensive irrigation, beginning about 1951, caused ground water tables to rise, creating many lakes and ponds (Figure 1.3). Some of these have sides too steep to support emergent vegetation (Figure 1.4), but where gradients are shallower there are stands of cattail (*Typha*), bulrushes (*Scirpus*) and sedges

FIGURE 1.2. Remnants of bunch grass (*Agropyron spicatum*) which formerly covered much of the uplands at the Potholes. It is now mostly replaced by the Old World annual, *Bromus tectorum*.

FIGURE 1.3. Lyle Lake, Columbia National Wildlife Refuge. The large stand of *Typha* supports territories of Redwings and Yellowheads. Sagebrush (*Artemisia tridentata*) and dry grassland can be seen on the lava hills in the background.

FIGURE 1.4. Pit-teal Lake, Columbia National Wildlife Refuge. This lake lacks emergent vegetation capable of supporting nests of blackbirds, and no Redwings or Yellowheads breed here. During most years there is a nesting colony of Brewer's Blackbirds in the sagebrush bushes on the hillside to the right of the lake.

13

(*Carex* spp.). Scattered stands of Peachleaf Willow (*Salix amygdaloides*), heavily damaged by beavers and cattle grazing, occur along streams and edges of some lakes.

The Turnbull National Wildlife Refuge, situated 30 kilometers south of the city of Spokane, is also underlain by lava which was scoured by Pleistocene floods. However, the topography is less steep, and all lakes are completely surrounded by stands of *Typha* and *Scirpus*. Due to its higher elevation (600 meters) precipitation is heavier (450 mm per annum), winters are colder, and the upland vegetation consists of an open parkland of Ponderosa Pine (*Pinus ponderosa*) with groves of Quaking Aspen (*Populus tremuloides*) in wet depressions (Figures 1.5 and 1.6). Water levels are regulated by structures at lake outlets, but levels drop dur-

FIGURE 1.5. Thirty Acre Meadow, Turnbull National Wildlife Refuge. The dense stand of *Typha* and *Scirpus* in the foreground is occupied by Redwings, while Yellowheads have their territories in the scattered clumps of emergent vegetation in deeper water. A grove of Quaking Aspen (*Populus tremuloides*), surrounded by a parkland of Ponderosa Pine (*Pinus ponderosa*), can be seen across the lake.

14

FIGURE 1.6. Little McDowell Lake, Turnbull National Wildlife Refuge. Yellowhead territories occupy the open vegetation in the deeper parts of the lake, while Redwing territories ring the periphery. Ponderosa Pine parkland covers the hill beyond the lake.

ing the summer, and some lakes regularly dry up completely in September except in unusually wet years.

Studies of Redwings in a humid environment were carried out in a marsh on Foster's Island in the arboretum of the University of Washington, Seattle. Marshes in this area are very dense, and many have a submerged mat of vegetation. On my study site there was such a mat at a depth of about 1 meter, generally sufficient to support my weight. The natural upland vegetation is a coniferous forest of Douglas Fir (*Pseudotsuga menziesii*), Western Hemlock (*Tsuga heterophylla*), and Western Red Cedar (*Thuja plicata*), but along the edges of wet areas Black Cottonwood (*Populus trichocarpa*), Red Alder (*Alnus rubra*) and willows (*Salix* spp.) are common. Except for lawns there is no open, herbaceous, upland vegetation. These marshes are marginal for blackbirds, and only Redwings breed there.

The Cariboo Parklands of British Columbia are also underlain by Miocene lava flows which form a gently undulating plain at an elevation of about 900 meters between the Coast Ranges and the Rocky Mountain complex (McKee, 1972). Upland vegetation consists of forests of Douglas Fir, Lodgepole Pine (*Pinus contorta*) and Quaking Aspen interspersed with grassland and spruce (*Picea nigra*) bogs. The lakes of this area vary widely in salinity, and the water levels of many of them fluctuate markedly from year to year. Blackbirds are, however, mostly restricted to lakes with fairly constant water levels because only there can they find emergent vegetation for nests. Winters are long and cold, but summers are warm with frequent showers. *Scirpus* dominates the emergent vegetation on most lakes (Figure 1.7), but there are patches of *Typha* and many sedges.

FIGURE 1.7. Rush Lake, Cariboo Parklands, British Columbia. The emergent stands of *Scirpus* are entirely occupied by Yellowheads except in one corner (not shown) where a woodlot is adjacent to the marsh. Redwings establish territories there.

More complete descriptions of these areas can be found in previous publications (Horn, 1968; Orians, 1966; Orians and Christman, 1968; Orians and Horn, 1969; Willson, 1966). The major features directly influencing the breeding biology of blackbirds are summarized in Table 1.3. Descriptions of study areas in California and Costa Rica,

TABLE 1.3. Characteristics of major study areas.

Lakes	Uplands	Blackbirds present
Seattle		
Dense vegetation, tendency to become boggy; low conductivity; very small emergence of aquatic insects; fish present; outlet and water level fairly constant.	Densely forested except where recently or continuously disturbed.	RW (BB only on lawns or golf courses)
CNWR		
Many, steep-sided; emergent vegetation only in small scattered patches, many low sedgy areas; conductivity uniformly high; large emergence of aquatic insects except in a few lakes which dry up and those infested with carp; trout in virtually all lakes; water levels mostly constant.	Sedgy swales, dry upland sagebrush and sparse grass; scattered willows along streams and occasionally by lakes; heavy cattle grazing.	RW YH BB
TNWR		
Mostly shallow with gradual sides; virtually solid bands of dense emergent vegetation around most lakes; water levels decline during spring and summer; great range of emergence rates from very low in acid lakes and ponds which dry up to very high in permanent, high conductivity lakes; no fish in most lakes.	Open Ponderosa Pine parkland with aspen groves in moist pockets; heavy cattle grazing but not until after blackbird breeding season.	RW YH BB (rare)

TABLE 1.3 (cont.).

	Lakes	Uplands	Blackbirds present
Cariboo Parklands	Lakes without outlets (and many fluctuate greatly in level in accordance with yearly rainfall); widely fluctuating lakes mostly lack emergent vegetation, others with more or less continuous bands of bulrushes; great ranges in conductivity and corresponding variation in quantity of emergence of aquatic insects; most lakes without fish.	Heavily grazed dry grassland and mixed pine-aspen groves; black spruce bogs.	RW YH BB
Costa Rica	Marshes form during rainy season and fluctuate according to amount of precipitation; dense growth of aquatic grasses and cattails during the rainy season; very little emergence of aquatic insects but dense populations of terrestrial insects on the grasses.	Naturally forested but pastures and rice fields border most marshes now; heavy cattle grazing and annual cutting of brush with machetes.	RW

which will be referred to briefly, are available elsewhere (Orians, 1961, 1973). The Argentina study areas will be described in detail in Chapter Seven.

1.3. MATERIALS AND METHODS

The project was carried out almost entirely in the field, employing standard observational techniques. The general behavior of most species was recorded and described in detail (Orians and Christman, 1968; Horn, 1970). This background information, not included except incidentally in

18

this monograph, was essential in permitting accurate interpretations of behavior observed in the field and was utilized in the study of spacing patterns and mate selection.

I made maps of most of the major study marshes on which I plotted territories of males by recording movements of individuals and noting locations of boundary displays. Because my study was extensive and involved a large number of study sites, I was unable to band or otherwise identify many birds individually, but most of the hypotheses being tested do not require knowledge of the behavior of individuals. For many territories, vegetation density was measured by means of a method similar to that used by MacArthur and MacArthur (1961). Every five paces (sometimes ten on marshes with large territories) measurements were taken of the distance from a point to the nearest new and old vegetation stalk in each hemisphere. In addition a measure was taken of the distance required to obscure with vegetation half of a 23.0 × 30.5 cm board painted white and hatched with a grid of black lines. Two such measures were taken at each point at right angles to the line of march.

Because of the importance of adjacent upland vegetation, measures were taken with a clinometer in many territories of the angle above horizontal subtended by the tallest upland vegetation or, as at the Potholes, of cliffs. Additional information about the position of the territory with respect to other territories, open water, and upland vegetation was provided by the territory maps.

Marshes were searched regularly for nests. On some of the more intensely studied areas these searches provide enough observations to permit an accurate estimate of the number of nests built in each territory and their success. Females spend much of their time low in the vegetation where they are hard to see and it is difficult to count them directly. Accordingly, I estimated harems by the minimum number of females that could account for the nests built on a territory. If a nest was started a week or more after destruction of another nest on the same territory, it was attributed to the same female that built the previous one. Dolbeer (1976) reported occasional renesting intervals as short as four days, but these were minimum estimates, and real values could have been five or six days (Dolbeer, personal communication).

Where possible, foraging of adult birds was observed directly. This was easiest to accomplish in the Potholes because of the openness of the vegetation and the presence of cliffs near most of the lakes. Foraging locations have also been inferred from food delivered to nestlings. Food samples were taken by putting segments of pipe cleaner around the necks of nestlings loosely enough to permit normal breathing but tightly enough to prevent passage of food items (Orians, 1966; Willson, 1966). A "sample" for the present purposes is the food obtained from all nestlings in a nest during a sampling period, usually about an hour in duration. During this study about 1,500 such food samples were obtained and analyzed. These data have been used for inferences about foraging habitats of the adults, relative rates of delivery of prey items, discriminations of prey, differences in foraging behavior among individuals, and seasonal and diurnal patterns of foraging. They cannot be used, however, to estimate absolute rates of delivery of prey because not all food items are retained by the pipe-cleaner collars, and the accumulation of food in the throats of nestlings doubtless affects begging behavior. Since only females bring food to nestling Redwings, a food sample represents foraging activities of a single individual. Male Yellowheads do feed nestlings, but most samples represent the foraging activities of just the female.

The availability of prey organisms was measured in several different ways. Emergence of aquatic insects was

20

measured at Turnbull in 1968 and in the Potholes in 1968, 1969 and 1970. At Turnbull ten traps, based on the design by Cook and Horn (1968), were established on each of six lakes. Of these traps five were shore traps to capture insects emerging on the edge (Figure 1.8), and five were traps placed out over emergent vegetation to capture insects crawling up stalks to metamorphose (Figure 1.9). At the Potholes traps were operated at four lakes in 1968, at three lakes in 1969, and at two lakes and along a stream in 1970. At one of the lakes, Coot Lake, which was studied all three years, there was no emergent vegetation, and only five shore traps were operated. On the other three lakes in 1968 and 1969 there were five shore traps and six floating traps. All traps were visited daily during the emergence period which began in late April. Normally all traps were checked late in the afternoon (Figure 1.10), but during June in the Potholes some traps were checked at dawn. When this was done damselflies, which emerge during the

FIGURE 1.8. Shore trap in position at Lyle Lake, Potholes. A dense bed of cattail separates the trap from the open water.

FIGURE 1.9. Floating traps being checked at Willow Pond, Potholes. The traps, open on the bottom, were mounted on 4 posts driven into the substrate far enough that the sides of the traps were just below the waterline. Emergent vegetation inside the traps was clipped frequently enough to prevent the traps from being lifted out of the water. A shore trap is being emptied at the far right.

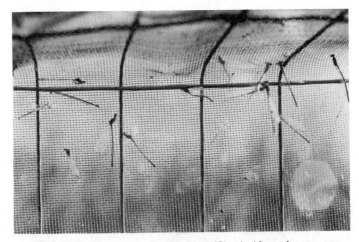

FIGURE 1.10. Recently emerged damselflies inside a shore trap, Hampton Lake, Potholes. Several exuviae can be seen attached to the screen. The teneral damselflies are capable of flight but normally do not attempt it if sunlight is no longer falling on the trap.

day, were assigned to the previous day while aeshnid dragonflies, which emerge at night, were assigned to the day of pickup. Daily trap checking was terminated at Turnbull at the end of July in 1968, and only sporadic checks were made for the remainder of the summer. At the Potholes in 1968 daily checks terminated July 16, and traps were operated only two days each week thereafter until early September. Since blackbird breeding terminates in late June in both areas, the sampling period included the critical period of utilization of this food source by the birds. In the uplands, samples were obtained with a standard sweep net in different habitats. At present I have no way of comparing these samples on the same scale with aquatic ones and cannot use them to estimate comparative prey availability, but the data are useful for assigning different prey species to habitats. This has been possible most readily in the Potholes and Cariboo Parklands where the habitats are simple.

In the laboratory, representatives of common prey items were burned in a Parr Oxygen Bomb Calorimeter. Once a representative set of caloric values was known, additional estimates were made on the basis of oven dry weight. This was used especially for species in groups such as spiders and grasshoppers in which there is a considerable range in size of individuals.

I took advantage of "natural experiments" provided by environmental heterogeneity. The most important of these were differences in productivity and patch types of different lakes. In addition, I utilized a number of changes, several of which constituted modifications which would have been difficult to accomplish by my own efforts. The most important of these changes was vegetation succession. Sometimes, I was provided with reverse succession in the form of opening up of a marsh by refuge personnel to improve its duck-producing capacity or the accomplishment of the same end by the combined activities of beaver and muskrats. The effects of fish on quantity of emergence of aquatic insects were demonstrated by the invasion of Carp (*Cyprinus carpio*) into several of my study lakes at the Potholes. All Potholes lakes I studied were stocked with trout, but most lakes at Turnbull lack fish.

The strategy employed in this study, like all conceivable strategies, imposes limits on the results obtained. For none of the major topics explored, such as foraging behavior, clutch size, mating patterns and habitat selection and spacing, are the results as complete as they would have been had any one of them formed the prime focus for an intensive investigation. Nevertheless, I believe that details which were sacrificed for the breadth of the study have been more than compensated by the insights derived from better knowledge of the relationships among them.

CHAPTER TWO

Marshes as Providers of Resources for Blackbirds

The important resources for breeding blackbirds provided by marshes have complex temporal and spatial patterns that strongly influence the best times and places for blackbirds to forage, establish territories, and build nests. The most striking characteristic of temperate marshes is the seasonal pattern of emergence of insects with aquatic larval but terrestrial adult stages. During the first field season of this study, we determined that these emerging insects were the most important component of the diets of breeding blackbirds in Washington and, consequently, great effort was expended in measuring the yearly, seasonal, and diurnal patterns of this emergence and in assessing differences among lakes and causes for them. This information is vital for testing theories of habitat selection, mate choice, foraging behavior and prey choice, and resulting patterns of niche breadth and overlap.

In addition, the emergent vegetation of marshes provides roosting cover, escape cover from predators during the day, and nest sites. Normally, such cover is not in short supply in relation to food resources, but in the more alkaline lakes, growth of emergent plants is depressed or prevented, and blackbirds are excluded from these lakes even though food supplies may be excellent. Also, the pattern of vegetation in marshes and adjacent uplands interacts with emergence of aquatic insects to determine encounter rates of foraging birds with their prey and, hence, suitability of various foraging sites and optimal choice of prey. Since prey availability is the most complex part of the marsh resource picture, this chapter will be devoted entirely to its description. Data on vegetation will be presented in subsequent chapters where appropriate.

2.1. QUANTITY OF EMERGENCE

The total number of aquatic insects available on different marshes is a function of the chemistry of the lake, its depth and permanence, kinds of predators present, and the nature of submerged and emergent vegetation. The numbers encountered at a specific site are strongly related to the position of that site within the vegetation matrix of the marsh. Our emergence sampling program was designed to measure the importance of all of these factors. General information on odonate emergence patterns is provided by Corbet (1962).

Conductivity of the water, a variable directly correlated with the concentration of dissolved salts and quantities of various dissolved materials, was measured for many of the study lakes because of its relationship to productivity. All of the Potholes lakes are formed by seepage from the Potholes Irrigation Canal or by a raising of ground water levels by irrigation. Therefore, water characteristics are uniform and the major differences among lakes relate to their depth and presence or absence of fish. The ages of the study lakes range from 10 to 30 years. At Turnbull and the Cariboo Parklands of British Columbia, however, the chemistry of the lakes also varies considerably, and in the Cariboo Parklands, the distribution of breeding blackbirds is more strongly correlated with lake chemistry, probably because of its influence on food production, than with the other factors (Orians, 1966).

The numbers of emergence traps and the periods of their operations on the various study lakes in 1968 are shown in Table 2.1. Traps were run continuously from just

26

TABLE 2.1. Number and period of continuous operation of emergence traps on the study lakes, 1968.

Lake	Type of trap	No. of traps	Period of operation
CNWR			
Coot	Shore	5	9 May—11 July
Willow	Shore	5	26 Apr.—11 July
	Floating	6	26 Apr.—11 July
Hampton	Shore	5	26 Apr.—11 July
	Floating	6	26 Apr.—11 July
Lyle	Shore	5	26 Apr.—11 July
	Floating	6	26 Apr.—11 July
TNWR			
Kepple	Shore	5	23 Apr.—31 July
	Floating	5	23 Apr.—25 June-31 July[a]
Beaver	Shore	5	23 Apr.—31 July
	Floating	5	23 Apr.—9-14 July [a]
Blackhorse	Shore	5	23 Apr.—31 July
	Floating	5	23 Apr.—11 June-12 July[a]
Mann	Shore	5	23 Apr.—31 July
	Floating	5	23 Apr.—1-10 July[a]
Big McDowell	Shore	5	23 Apr.—31 July
	Floating	5	23 Apr.—31 July
Little McDowell	Shore	5	23 Apr.—31 July
	Floating	5	23 Apr.—15 June-28 July [a]

[a] Dates indicate span of termination period for the operation of floating traps due to falling water levels, e.g. at Kepple the first trap failed on 25 June, the last on 31 July.

prior to the beginning of emergence until after the blackbirds stopped breeding. At the Potholes, sampling was continued beyond the blackbird breeding season on an irregular basis, but earlier termination of some traps at Turnbull was dictated by falling water levels. The seasonal totals of insects captured, and rates of emergence on all of the lakes at which measurements were taken at the Potholes and Turnbull are shown in Table 2.2. At Turnbull, water levels dropped sufficiently on some of the study lakes so that all emergent vegetation was above the waterline by late July, and only shore traps, which were daily moved to remain with the receding water, were in a position to capture insects.

The permanence of a pond or lake exerts a major influ-

TABLE 2.2 Emergence rates of odonates as measured by traps at the Columbia and Turnbull National Wildlife Refuges, 1968.

| | | Daily emergence of odonates: | |
Lake	Total no. of odonates captured	per meter of edge	per square meter of emergent veg.
CNWR			
Coot	11,464	47	—
Willow	1,583	2	4
North Hampton	11,039	9	36
Lyle	249	0.5	0.5
TNWR			
Kepple	17,600	31	38
Beaver	282	0.6	0.2
Blackhorse	5,727	9	8
Mann	71	0.2	<0.05
Big McDowell	12,190	22	9
Little McDowell	3,565	5	7

ence on the kinds and numbers of insects emerging from it. Most of the common odonates of permanent lakes in Washington overwinter in late instars, and these species are unable to live in lakes that dry up during the late summer, the dry period in Washington. These lakes, which include Beaver Pond and Mann Lake at Turnbull and Willow Pond at the Potholes, support only populations of species that overwinter as eggs and complete their life cycles during spring and summer. Emergence rates were much lower on these three than on other lakes at those areas (Table 2.2 and Figure 2.1), and the peaks came later in summer, mostly after the blackbird breeding season.

None of the major study lakes at Turnbull contain fish, and the major predators on odonates in these lakes are probably larvae of Tiger Salamanders (*Ambystoma tigrinum*) and larvae of predatory beetles (especially Dytiscidae). Most Potholes lakes that hold permanent water are stocked with trout. The fact that these lakes, despite their favorable chemistry, have lower emergence rates than those at Turnbull may reflect the influence of trout on insect populations, but because of other differences a precise estimate of the effects of trout cannot be made.

Much more striking, however, are the effects of carp. Carp have entered the Potholes drainage systems from the Columbia River. Of the lakes studied at the Potholes, only Coot Lake, Willow Pond and North Hampton were free of carp during the entire study. The reduction of emergence of insects by carp is dramatically shown by the emergence at Lyle Lake during 1968, a year when it had extremely high carp populations. No emergence data are available for earlier years, but in 1963 when I began studies at the Potholes, Lyle Lake was an excellent insect-producing lake and supported dense breeding populations of blackbirds. The virtual elimination of breeding blackbirds as a result of the carp invasion will be documented later. The emergence at Lyle Lake was reduced by carp to about 1/100 of its probable previous value, primarily because of habitat destruction rather than direct predation on insects. Carp-infested lakes are characterized by muddy bottoms almost devoid of vegetation, and their water is perpetually murky. Carp-free lakes, on the other hand, have clear water and an abundance of submerged vegetation which provides shelter and foraging perches for insects. The only odonates to survive in carp-infested lakes are those living at the edge where water is too shallow for carp to forage.

In the absence of carp, larval odonates are found throughout the lake, on the bottom and foraging from submerged vegetation, depending on the species. At the time of emergence they move toward shore and emerge either on the shore itself or on emergent vegetation. With the exception of dragonflies of the family Libellulidae, which swim to the shore to metamorphose, all odonates in Pacific Northwest ponds tend to crawl out on the first emergent substrates they encounter in their movement from deeper water toward shore. If there is no band of cat-

(a)

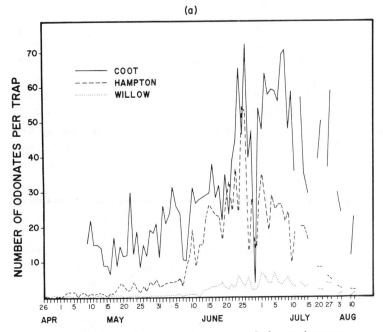

FIGURE 2.1. Seasonal patterns of emergence of odonates in eastern Washington. (a) daily captures per trap at three Potholes lakes in 1968; (b) (facing page) daily captures per trap, averaged weekly, of emerging odonates at four Turnbull lakes in 1968.

tails, bulrushes, or other emergent vegetation, the larvae swim all the way to the shore and emerge there, but if there is a band of emergent vegetation they usually crawl up the outer stalks (Figure 2.2). The effects of vegetational geometry on emergence rates at different sites are shown in Figure 2.3. Shore traps at Hampton Lake and Kepple Lake caught six to fifteen times as many odonates if they were exposed to open water than if they were located behind beds of cattails and bulrushes. At Hampton, traps at the outer edge of the cattails caught three times as many insects as those situated well within the vegetation even though cattail beds on that lake are not very dense. At Kepple Lake an edge effect was not noted, but the results

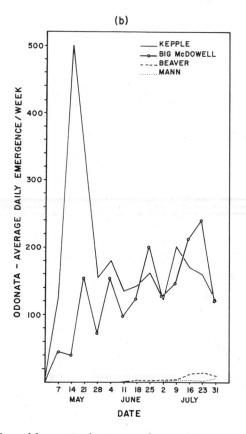

are equivocal because the vegetation was open and patchy in the sampling area and because all floating traps were useless by June 25 because of dropping water levels.

In general, then, the best foraging areas for blackbirds are outer edges of patches of emergent vegetation and shores of lakes at locations where there is no offshore emergent vegetation. Shores behind emergent vegetation and interiors of patches of emergent vegetation are usually much poorer foraging areas. Though I do not have direct data, the interiors of extensive beds of cattails and bulrushes are probably poorer than those areas we have

FIGURE 2.2. Larval exuviae of odonates on a bulrush stalk, near Phalarope Lake, Cariboo Parklands, British Columbia. Normally they are quickly blown or washed into the water.

studied. The broader the expanse of emergent vegetation, the fewer insects should emerge per unit area of vegetation, unless there is a substantial emergence due to insects feeding within the vegetation. Production of odonates within cattail beds is probably very low because emergent plants interrupt most of the sunlight, and little penetrates into the water to support an *in situ* food chain capable of sustaining large populations of predatory insects. The na-

□ KEPPLE

■ HAMPTON

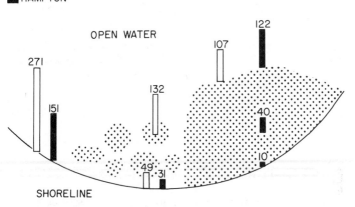

FIGURE 2.3. Effects of trap location on quantity of emergence of odonates at Kepple Lake, TNWR, and Hampton Lake, CNWR in 1968. Columns represent average odonates trap^{-1} week^{-1} for floating and shore traps in the situations indicated. Stippled areas represent patches of emergent aquatic vegetation.

ture of the detritivore-based food chain is poorly understood, but it clearly does not result in large emergences of prey suitable for exploitation by blackbirds.

Marshes also support insects with strictly terrestrial life cycles, such as Lepidoptera, Hemiptera, Diptera and Orthoptera, which eat leaves of emergent aquatic vegetation (Claassen, 1921). Our data indicate that they are of minor importance to breeding blackbirds in the Pacific Northwest. They were never common during our studies and do not reach their peak populations until late in summer, long after blackbirds have stopped breeding. Also, some lepidopterans and hemipterans are conspicuously colored and have "hairy" surfaces; strong indicators that they are toxic or distasteful (Eisner, 1970; Rodriguez and Levin, 1976; Root, 1966). We have recovered none of them from any blackbird food samples and have, accordingly, not attempted to measure their abundance.

33

On the other hand, predatory arthropods, such as spiders, are significant and, in some situations, major components of the diet. Our sampling methods do not capture them, however, and we have no estimates of their abundances. Nor have we estimated abundances of insects, such as Dytiscidae and Notonectidae, that have strictly aquatic life cycles. Some inferences about their availability can be drawn from blackbird foraging data, but the conclusions must be more tentative than for prey that have been sampled intensively.

2.2. TEMPORAL PATTERN OF EMERGENCE

On all study lakes except Kepple, emergence of odonates reached a peak in late June and continued at fairly high levels through July (Figure 2.1a,b). At Kepple, in 1968, emergence peaked in May. Odonates dominated the emergence numerically, and because they are also the largest emerging insects, they are the most important food source for breeding blackbirds.

The emergence pattern of dipterans, especially chironomids, the second most important group of insects, was markedly different, peaking in late May and dropping to very low levels by late June (Figure 2.4). Ephemeroptera, the only other group of insects emerging in significant numbers, had a more irregular pattern, with peak periods in mid to late May and again in late June and early July (Figure 2.5). In comparison with odonates and diptera, however, their numbers and biomass were trivial.

At the Seattle study marsh, there was almost no emergence of odonates or Diptera, and these insects were always relatively scarce. Beetles of the genus *Donacia* (Chrysomelidae), whose larvae eat roots of aquatic plants such as cattails and bulrushes, were a major food item for nestling Redwings, and we sampled their abundance in cattail stalks

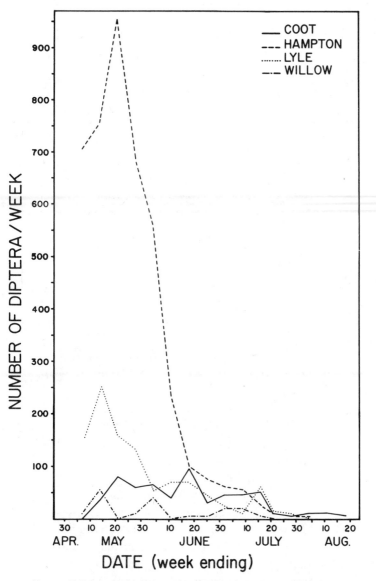

FIGURE 2.4. Seasonal pattern of emergence of Diptera from four Potholes lakes in 1968. Curves plot weekly averages per trap.

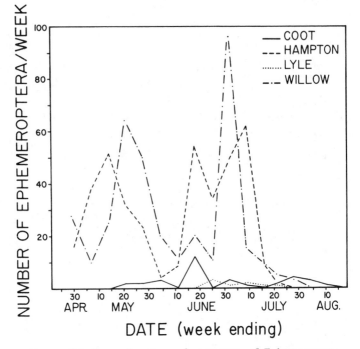

FIGURE 2.5. Seasonal pattern of emergence of Ephemeroptera from four Potholes lakes in 1968. Curves plot weekly averages per trap.

where they hide. Weekly, beginning early in February and continuing until early June, 100 stalks of cattails were searched for beetles. Their abundance peaked in late April and dropped sharply immediately thereafter (Figure 2.6).

Diurnal patterns of emergence of aquatic insects were measured at Kepple and Blackhorse lakes at Turnbull and at Coot and Hampton lakes in the Potholes. Traps were visited every two hours during the day, beginning at 0600 or 0800 hours, and all insects were removed and counted. Since total emergences on days when 2-hourly samples were taken were not lower than on days when the traps were visited only in the evening, our activity did not appear to discourage emergence during succeeding time intervals.

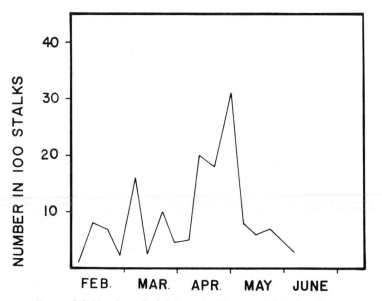

FIGURE 2.6. Number of adult *Donacia* (Chrysomelidae) found per 100 stalks of cattails, Foster's Island, Seattle, 1968. Sampling was done weekly from early February through early June, covering the entire Redwing breeding season.

On all lakes there was a marked peak of emergence of damselflies during the late morning hours (Figure 2.7). Since collection hours and not emergence times are plotted in this figure, the actual emergence pattern was advanced about one hour. The late morning peak is most pronounced early in the season when nighttime temperatures are much colder. This pattern is probably adaptive because emergence is concentrated during a period when metamorphosis can take place most rapidly, thereby minimizing the period of extreme vulnerability to predators. Strong winds, which are common in the afternoon, especially during the earlier part of the season, may contribute to making afternoon emergence more risky even though temperatures are favorable. As soon as a teneral odonate has hardened sufficiently to fly, it leaves the lake and completes its

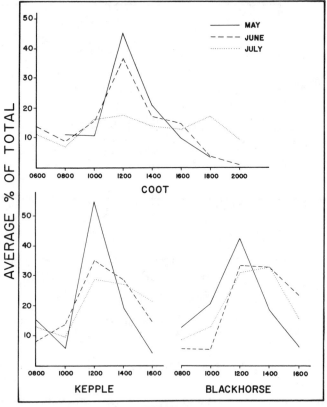

FIGURE 2.7. Diurnal patterns of emergence of odonates at one Potholes lake (Coot) and two Turnbull lakes (Kepple and Blackhorse) in 1968. In all three lakes mid-morning synchrony of emergence is greatest early in the season.

hardening and attainment of color elsewhere. There is accordingly a major increase in abundances of these insects in sagebrush in the Potholes and in grass, shrubs, and trees at Turnbull during the afternoon. At this time, tenerals are still easy prey for blackbirds. The fully-hardened adults, however, are difficult to capture and are taken mostly very early and late during the day when their flight activity is inhibited by low light levels and temperature.

The quantity and pattern of emergence is strongly influenced by the weather. Heavy cloud cover and strong winds inhibit emergence, and on days of heavy rain emergence may be completely absent. Emergence data from Turnbull for 1968 have been plotted according to weather conditions in Table 2.3. To correct for seasonal changes in emergence rates, each day is compared with the mean for the surrounding week. On clear to partly cloudy days without rain, emergence is usually within 20% of the weekly mean, but emergence was reduced on days with bad weather; two-thirds of the rainy days had emergences of less than half the weekly mean. Morning clouds and rain may also delay emergence several hours. On some days at Turnbull, when there was morning rain and afternoon clearing, the emergence peak occurred as late as 1500 hours.

Diurnal patterns of emergence also occur in other insect groups, but our traps do not sample them as effectively. Midges, for example, emerge over the entire surfaces of the lakes and are not concentrated by the location of emergent vegetation. Though we collected most midges on the first morning visit to the traps, most of them probably emerged at dusk after the traps had been emptied (Figure 2.8). Trichopterans also emerge primarily at dusk, but our traps captured very few of these insects, because they also emerge directly at the water surface.

An additional source of temporal variation appears to be damselfly life cycles. Early in the season on both Coot and McDowell lakes, there was a cycle of emergence not related to weather conditions. These variations may represent emergence of successive groups of individuals that overwintered in the ultimate, penultimate and antepenultimate instars, respectively (Corbet, 1957). If development were

TABLE 2.3. Effects of cloud cover on the emergence of odonates at the Turnbull National Wildlife Refuge, 1968. Emergences on each day are compared with the mean daily emergence during the week in which the day occurred. The numbers of days with rain are indicated in parentheses.

Cloud cover	No. of days on which the emergence rate was:				
	>50% less than weekly mean	20-50% less than weekly mean	Within 20% of weekly mean	20-50% greater than weekly mean	>50% greater than weekly mean
Completely overcast	4 (3)	1	3 (2)	0	0
Mostly overcast	1 (1)	2	8 (1)	1	0
Partly cloudy	0	1	12	2	1
Mostly fair	0	1	9	3	1
Fair	1	3	24	4	2

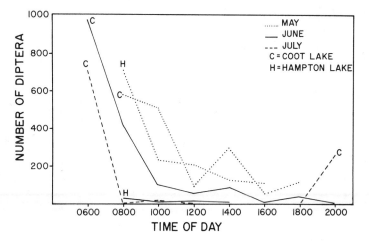

FIGURE 2.8. Diurnal pattern of emergence of Diptera at Coot and Hampton lakes, Potholes, in 1968. Plotted are total emergences from 5 traps at Coot and 11 at Hampton.

triggered by rising water temperatures, as seems likely, such a pattern could easily result.

For reasons not investigated by us, but probably related to events involving conditions for the larvae, there are important yearly differences in the pattern and quantity of emergence. Spring temperatures exert some influence on initiation of emergence, but other effects are even more powerful. Comparative data, available for Coot Lake in 1968, 1969 and 1970, and for Hampton Lake in 1968 and 1969 (Figure 2.9), show that blackbirds have difficulty predicting both the timing and the amount of emergence, but we have insufficient data to compare this uncertainty among regions, lakes, or sites. Nor do we know what kinds of clues might be available to birds when they must make their decisions. At our study sites males select territories two to three months prior to the beginning of emergence, females about one month in advance.

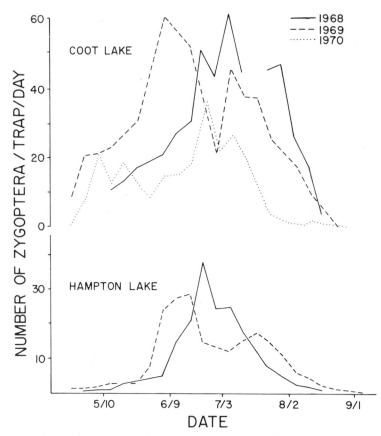

FIGURE 2.9. Yearly differences in pattern and quantity of emergences of damselflies at two Potholes lakes.

2.3. ABUNDANCE OF INSECTS ON THE UPLANDS

Insect abundance on the uplands surrounding a marsh reflects (a) production on the areas themselves and (b) movement of aquatic insects from the water after metamorphosis. Since primary production is strongly correlated with evapotranspiration (Holdridge, 1957; Rosenzweig, 1968), insect abundance in upland habitats should be

42

higher in regions with higher rainfall, but the contribution of aquatic insects to total abundance depends on productivity of local lakes. These two sources of insects are, in a broad geographical sense, inversely correlated. Quantities of dissolved nutrients are usually higher in lakes in arid and semiarid regions (Rawson, 1961; Edmondson, 1963) where upland productivity is low. Conversely, lakes of more humid regions regularly flush out and do not concentrate nutrients. They produce fewer aquatic insects, but productivity is higher on adjacent uplands.

At the Potholes, upland sagebrush and dry grass support low numbers of orthopterans, leafhoppers and lepidopteran larvae (especially Pterophoridae), and the densities of breeding birds that depend on these resources, Western Meadowlarks (*Sturnella neglecta*), Sage Sparrows (*Amphispiza belli*), and Lark Sparrows (*Chondestes gramminicus*), are very low. During spring and summer afternoons, however, the bushes often swarm with midges, damselflies, mayflies and caddisflies. At this time blackbirds forage extensively in the uplands (Orians and Horn, 1969).

At Turnbull, vegetation is more complex, and insect communities are more difficult to characterize. In 1968, sampling was done on eight different days with 25 sweeps of a standard sweep net in Canary Grass (*Phalaris arundinacea*) along marsh edges, in shrubs, especially Snowberry (*Symphoricarpus albus*) and roses (*Rosa* spp.) about 30 meters from the shore, in upland grasses and forbs about 65 meters from the shore, and in herbs and shrubs in an aspen grove about 95 meters from the shores of Blackhorse Lake (Figure 2.10). The number of arthropods greater than 4 mm in length reached a peak in early to mid-May along the shore and in the upland shrubs and grass. Comparable sampling was not done in the aspen grove, but in early June insects were more abundant there than elsewhere and more abundant than at any time dur-

43

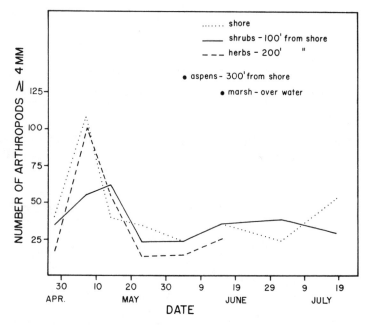

FIGURE 2.10. Number of arthropods ≥ 4 mm in total length captured in 25 sweeps of a standard insect beating net in various terrestrial patch types at Turnbull, 1968. For comparison single samples from foliage of Quaking Aspen 300 ft from a marsh and from emergent vegetation over water are also shown.

ing the sampling period at other sites. For comparison, a single sample from aquatic vegetation, taken in early June, is presented to show the abundance of insects there. At all distances from the lake, aquatic insects constituted a significant portion of the total, especially in May. The high figures for the aspen grove may in part be due to indigenous aquatic insects because the groves are in standing water in the spring and early summer.

At the Costa Rican study site, emergence of aquatic insects was relatively small and consisted primarily of dragonflies which emerged at night and were, therefore, largely unavailable to blackbirds. However, the marshes were choked with a variety of herbaceous plants, some of which supported large populations of lepidopteran larvae and orthopterans, the principal prey of the breeding blackbirds (Orians, 1973). These insects have marked activity changes between day and night, but during daylight hours when blackbirds are active, there appeared to be no significant changes in their locations and availability.

The picture of emergence in the Pacific Northwest provided by these data can be summarized as follows:

PATTERN OF EMERGENCE:

(a) Emergence of aquatic insects begins in late April, reaches a peak in late June, and continues through the summer.

(b) During a typical day insects begin emerging early in the morning, reach a peak just prior to noon, and decline rapidly in the afternoon. This diurnal pattern is more marked earlier in the season than later.

(c) Daily emergence is more predictable in June and July than in May.

(d) The movement of recently emerged insects away from the edges of lakes and ponds greatly increases prey density in the uplands during the afternoon.

(e) Within lakes, emergence is greatest on shores exposed to open water and on outer edges of patches of emergent vegetation, and least on shores behind emergent vegetation and within expanses of emergent vegetation.

QUANTITY OF EMERGENCE:

The density of emerging aquatic insects in a lake is increased by:

(f) Available nutrients in the water.

45

(g) Permanence of water throughout the year.
(h) Absence of fish, especially carp, which have a catastrophic effect on emergence.

There is reason to believe that most of these patterns are characteristic of many marshes of arid and semiarid regions of western North America. Though we did not measure emergences with traps in British Columbia, the late morning peak of damselfly emergence is readily apparent to anyone wading through the marshes. I also made similar observations on insects in marshes in Utah and California. Evidence from the literature indicates that emergence rates are probably much lower in more humid regions where lakes do not concentrate nutrients (Bick and Bick, 1965; Byers, 1940; Corbet, 1952; Kormondy and Gower, 1965; Lutz, 1968; Pajunen, 1962).

It is significant that the quantity of emergence of aquatic insects in the Pacific Northwest is generally predictable from a knowledge of the date, time of day, and weather conditions. Therefore it is potentially possible for birds to "anticipate" future prey availability and to make decisions concerning settling location, foraging patches, and prey choices without having to sample them directly. This is especially important for territory establishment, which may occur more than a month in advance of the beginning of emergence. Considerable improvement in rate of capture of prey during foraging bouts could result from an ability to determine good foraging sites without having to visit many different areas. Accordingly, a sophisticated system of habitat assessment and rapid responses to temporal patterns of prey availability are to be expected among marsh-nesting blackbirds. The extreme variability of marshes makes them excellent environments in which to observe changes in bird behavior and to test various theories of responses of individuals to a fluctuating environment. For reference purposes, characteristics of the major study lakes are provided in Table 2.4.

TABLE 2.4. Summary of characteristics of study lakes and marshes.

Lake	Edge vegetation	Emergence characteristics
CNWR		
Coot	Sedges only	High emergence
Hampton	Mostly cattail & bulrush beds, some sedge	Moderately high emergence
Willow	Dense cattails	Low emergence; dries up in late summer in dry years
Lyle	Some cattails, lots of sedge	Initially high emergence but drastically reduced by carp
Pit	Some cattails, lots of sedge	Initially moderate, greatly reduced by carp
TNWR		
Kepple	Some cattail & bulrush, lots of sedge	Very high emergence
Beaver	Dense cattail	Low emergence, often dries up
30 Acre	Beds of cattails & bulrushes	Low to moderate emergence
Blackhorse	Beds of cattails & bulrushes	Low to moderate emergence
Mann	Mostly bulrushes	Very low emergence, often dries up in late summer
Little McDowell	Mixed cattails & bulrushes	Moderately good emergence
Big McDowell	Cattail, bulrushes & sedges	Moderately good emergence
Lower Turnbull	Extensive beds of bulrushes	Good emergence most years
Cariboo Parklands		
Rush	Pure stands of bulrushes	High emergence
Westwick	Pure stands of bulrushes	Very low emergence
Seattle		
Foster's Island	Dense stands of cattails, bordered by woods	Almost no emergence

CHAPTER THREE

The Adaptations: Selection of Habitats, Territories and Mates

The important decisions organisms make concerning where they will perform various activities are usefully viewed as a sequence from general to specific evaluations. First, an organism selects a habitat, a collection of patches of sufficient size and variety that it can carry out all of its activities for a breeding cycle, or lifetime, within its confines. The quality of a habitat is thus a function of the kinds and number of patches within it and their positions relative to one another.

Once a habitat has been selected, most organisms make many choices about the patches they will utilize, the temporal pattern of this use, and activities to be carried out within different patch types. Patches can be expected to differ in their suitability for different activities, that is, the best patch for avoiding predators may not be the best feeding site, and so forth. A number of problems relating to the use of patches for foraging will be dealt with in the following chapter. Here my focus is on selection of an overall habitat for breeding when reproduction requires exploitation of the environment from a central place for a long period of time.

In general, decisions about habitats are made less often than decisions about patches, which are, in turn, made less often than decisions about individual prey items. However, a habitat decision commits an individual to a physical location for an entire breeding season and, hence, it constrains and influences a large number of patch decisions and an even larger number of prey decisions. Therefore, a habitat decision can be expected to influence fitness more than a decision about patches or prey. Selection of a breeding site raises different problems for males and females of many species. Among the blackbirds I have studied, Brewer's do not defend a breeding territory, and males and females arrive on the breeding grounds already paired. Little is known about processes of mate selection in this species nor has the process of selection of nest site been analyzed in detail, though factors influencing colony location are reasonably well known (Horn, 1968; Furrer, 1974). Among Redwings and Yellowheads, as males arrive first and establish territories, females are able to make simultaneous assessments of the quality of a male and of his real estate. Therefore, for a female, selection of a mate is nearly equivalent to selection of a habitat and territory. Males, of course, make habitat decisions independently of the presence of females. This chapter is concerned primarily with selection of habitat and mates by Redwings and Yellowheads.

In these two species territories provide nest sites and some, but not all, of the food consumed by the adults and their offspring. Adults also roost on their territories during most of the breeding season. The intrinsic value of a territory should therefore be determined primarily by its food-producing capacity and the availability of safe nest and roost sites. All three of these factors are the result of combined values of many patches within the territory, and all of them are likely to be influenced by population density. The probability of arriving at a given territory is determined by the length of time over which an individual searches, the rate of its movement while searching, and its mortality rate while searching. Birds in general can move efficiently at high speeds when searching for habitats so that the probability of finding the best site in a general area is probably normally very good. However, the time available for searching may be short because the best territories are occupied quickly, and later arriving individuals must either displace earlier ones or accept poorer-quality sites. A bird that has located a breeding habitat of intermediate quality may select one of a number of options. First, it may settle in that habitat. Second, it may reject that habitat to search for a better one. A better site, if found, may be occupied or unoccupied, at which time the individual may continue searching, may attempt to evict one of the prior residents or to insert itself among them. Finally, the individual may remain in the vicinity of a good site waiting for the death of one of the residents. The expected payoff from these various options is a function of the species, circumstances, and time of the breeding period. The most useful theories concerning these behaviors attempt to predict behavior as a function of the consequences of different choices under varying circumstances.

In this chapter I consider theoretical predictions about processes of habitat selection by males and females, and present some field tests of them. Then I turn to the question of the factors influencing sizes of territories and examine patterns of variability in sizes of territories of blackbirds. The clues available to individuals to select mates and territories and how they are used is the focus of the subsequent section. Finally, I synthesize my observations to suggest the general causes of the patterns of dispersion of blackbirds in their breeding marshes.

3.1. HABITAT SELECTION BY MALES

Among birds, males are potentially capable of engaging in all types of parental investment during the breeding season except making and laying eggs. This single difference is an important one, however, because of the great differences in the energy contents of eggs and sperm (Orians, 1969). The reproductive output of a male is enhanced by mating with more than one female whereas that of a female usually is not enhanced by mating with more than one male. As a result, there is greater variance in breeding success of males than of females, and males compete for females among most birds (Trivers, 1972). The hypotheses and predictions to follow are based on a system in which males establish territories prior to the arrival of females and in which the possibility of attracting more than one female exists.

An abundance of scattered field data from many species of birds indicates that an attempt to evict an established resident from its habitat is not normally successful. This gives strong priority to the earliest possible settling in habitats that are only seasonally available, probably accounting for the very early spring migrations of most temperate zone birds as compared to the fall migrations, which usually take place when temperatures are still much milder than they are during the spring migration of the same species (Lack, 1965). In addition, mortality rates of birds during the breeding season are so low that there is a very low probability that a territorial holder will die soon enough during the breeding season that a replacement individual could reproduce during the same season. Therefore, among temperate-zone birds we expect extensive searching and a willingness to settle in suboptimal habitats. At lower latitudes, however, where breeding seasons are longer and habitat occupancy is often more permanent, the chances of finding unoccupied habitats, even with extensive search, are lower. Accordingly, habitat selection patterns there may be very different. The higher incidence of flock territories and extra helpers at the nest may reflect this selective shift (Fry, 1972; Skutch, 1935; Orians, Orians and Orians, 1977; Woolfenden, 1975).

The defense of any piece of ground carries with it a cost, in terms of time and energy devoted to defense, plus a risk, in terms of increased predation rates resulting from behaving in ways which both attract predators and reduce awareness of their presence. The benefits which compensate for these costs lie in retention of resources for exclusive use and/or reduction of rates of predation or parasitism by making offspring less likely to be discovered by a predator (Crook, 1964; Horn, 1968; Tinbergen, Impekoven and Franck, 1967).

For any specific level of food abundance and territory size, costs of defense and traveling time when the young are being fed are minimized if the territory is roughly circular and if the nest is centrally located. Most published observations of territory shape and nest locations in birds are in accordance with this prediction (Grant, 1968; Harris, 1944; Drury, 1961; Nice, 1943; Pitelka, Tomich and Treichel, 1955; Cade, 1960; Cody, 1969). The optimal location of a nest, whether or not the species is territorial, is that which minimizes total travel time during the period when food is being brought to the nest (Horn, 1968). This is a function of the spatial and temporal distribution of the food resources. The greater the variability of food distribution in space and time, the larger the area over which foraging must take place to capture an equivalent amount of food. Therefore, the rate at which energy can be delivered to the nest deteriorates as distributional variability increases, and available food resources are not economically defendable (Brown, 1964). With more strongly nonuniform food distributions, moreover, the advantages of being able to observe the foraging successes of others increases. Therefore, the presence of nearby individuals may enhance rather than decrease rates of delivery of food to the nest.

Territorial defense mechanisms should evolve to reflect the general level of food availability and may not reflect yearly fluctuations, unless clues to future food abundance are already available when territories are established. Sometimes this is true, but often it is not. Therefore, attempts to correlate territory size with yearly fluctuations in food availability are seldom successful. However, if food availability is strongly correlated with a pattern of environmental patchiness, as should often be the case, predictable relationships ought to exist between habitat quality and territory size.

The value of possession of an exclusive area in polygynous species like Redwings and Yellowheads is enhanced by the possibility of attracting multiple females. This should favor taking more risks in defense of space, such as arriving earlier in spring when feeding conditions may be poorer, investing so heavily in defense that weight drops during the defense period, and, on a longer-term basis, evolution of physical features that increase competitive fighting ability at the expense of survival under other circumstances (Maynard Smith, 1958; Hamilton, 1961; Orians, 1969). For example, territorial male Redwings and Yellowheads regularly lose substantial amounts of weight during the breeding season while first year (nonterritorial) males do not (Orians, unpublished data; Searcy, 1977).

3.2. HABITAT SELECTION BY FEMALES

Because of their energetically expensive gametes, females are normally limited in their reproductive output by the amount of energy that they can mobilize and not by the number of males with which they can mate. If males provide nothing but a set of genetic instructions, females should select their mates by assessing the probability that offspring resulting from a mating with a particular male will survive to reproductive age and their success when they do reproduce. If the male also provides parental care, the above probabilities will be influenced by the form and amount of his investment. Some forms of male parental care, such as providing a suitable territory and antipredator behavior, can be performed for the offspring of more than one female with little diminution of effectiveness, while others, such as providing food for offspring, are inherently indivisible. However, if nests of different females on a male's territory are staggered, the male may be able to provide food successively to nestlings of all females (Verner, 1964).

Since the quality of male parental care that a female can expect to receive is influenced by the presence of other females and the timing of their reproduction, females should be influenced in their choice of nesting sites by the presence of others in the territory and should attempt to modify the pattern of settling of additional females. Preventing other females from settling may be relatively easy while a female is nest building and laying eggs but, once incubation has begun, chasing a new female requires leaving the nest, which risks loss of the entire clutch by chilling and leaves the eggs unguarded. However, overlap in time of feeding offspring, the period when resources and male parental care may be most critical, is reduced if successive females do not initiate nests within two weeks of previous females. Therefore, the value of preventing other females from settling is relatively weak once a female has begun incubating.

In Redwings and Yellowheads, both of which have relatively large populations of nonterritorial males during the breeding season, I have no evidence of nonbreeding females except for one year in Costa Rica when drought and marsh drainage greatly reduced the amount of available nesting habitat in my study area (Orians, 1973). This suggests that all females capable of breeding settle somewhere, and defensive behavior by females does not influence the total breeding population though it may affect local distributions and timing of initiation of nesting by some individuals (see also Holcomb, 1974).

By influencing settling of additional females, a female can presumably affect the expected parental investment by her mate. For example, if the probability of attracting additional mates is high, a male may improve his fitness by investing all his time attempting to attract additional mates instead of investing in the offspring he already has. However, if the probability of attracting additional mates is low, whether or not this is influenced by the behavior of already settled females, then a male is more likely to gain by investing more in existing offspring, especially since their reproductive value increases with their age (Medawar, 1957).

In a very provocative and stimulating paper, Trivers (1972) suggested that an individual should be tempted to terminate investment in a set of offspring if it had invested less than its spouse. This is probably incorrect, because a decision to cease investing should be based on the expected contributions to fitness of continuing to invest compared to stopping (Dawkins and Carlisle, 1976; Boucher, 1977). The amount of past investment by either spouse is irrelevant to this decision *unless* it influences future costs and benefits of the options, as is the case with investment theory in economics where the amount of past investment is irrelevant to decisions about future investments (Alchian and Allen, 1964, chapters 15 and 29; Samuelson, 1970, chapters 11 and 30).

Past investments are most likely to influence future benefits by altering the probability that the spouse will continue to invest if its mate stops and by affecting the probability that another mate can be obtained. The relative cheapness of sperm enhances possibilities for additional matings, whereas the high cost of eggs makes production of a new clutch much more difficult. Nevertheless, for many birds, a deserting individual may have a low probability of finding an additional mate, regardless of past investments, if individuals of that sex are not available in greater numbers than those of the deserting sex. Therefore, I have made my predictions on the basis of probabilities of attracting and engaging in successful reproduction with additional mates rather than considering past investments of members of a pair when a decision to terminate investment is made.

3.3. PREDICTIONS AND TESTS ABOUT HABITAT AND MATE SELECTION

All decisions about habitats and mates are made with incomplete information, but it is nonetheless useful to develop an approximate theory by first assuming that the individuals have perfect knowledge. Once appropriate decisions are known for these conditions one can then consider how seriously in error decisions will be if made with different levels of ignorance. This provides a basis for considering how much individuals should invest in improving their knowledge before making decisions.

In the models that I have tested, an organism is presumed to make decisions concerning habitats in such a manner that its total fitness over the year is maximized. Thus, an organism should compare probable reproductive success and its own survival in the various habitats that might be encountered, appropriately discounted by the probability that they will be found and can be occupied if found. If there are no social constraints on habitat selection and each individual is able, with perfect knowledge, to select the best available habitat at the time it settles, then we expect a population to assume an *Ideal Free Distribution* (Fretwell and Lucas, 1969) in which different settling densities exactly compensate for intrinsic differences in habitat qualities (Figure 3.1). The effect of imperfect knowledge is

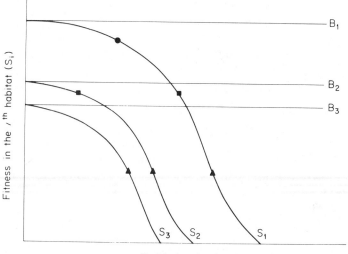

FIGURE 3.1. Graphical representation of an Ideal Free Distribution. Fitness is assumed to decrease with increasing population densities in all habitats. B_1, B_2, and B_3 represent maximal fitness in habitats S_1, S_2, and S_3 respectively. Individuals should not settle in S_2 until fitness in S_1 has been reduced to B_2. Similarly, individuals should not settle in S_3 until fitnesses in S_1 and S_2 have been reduced to B_3. The circle represents a possible final settling density at a low population level. The squares represent a possible final density when not enough individuals are present to cause any settling in S_3. The triangles represent a possible final settling density when all three habitats are occupied. In all cases differences in densities in the different habitats exactly compensate for initial differences in habitat quality (modified from Fretwell and Lucas, 1969).

to introduce variations in mean fitnesses in different habitats.

If, on the other hand, individuals are prevented from settling in some habitats by resistance from individuals already there, then average reproductive success should consistently be better in good habitats than in poorer ones, giving rise to a *Dominance Distribution* (Fretwell and Lucas, 1969). If the settling of additional individuals lowers mean success of individuals already present, benefits always accrue from excluding other individuals, but exclusion should not occur unless costs are less than benefits. For the reasons indicated above, the ability of a female to prevent settling of other females depends strongly on the stage of nesting cycle, but males are capable of exerting defense throughout the season at a relatively constant cost. Therefore, among Redwings and Yellowheads, I expected males to exhibit a strong dominance distribution whereas females would approximate an ideal free distribution.

For both sexes, the first individuals to arrive on the breeding grounds should select the best habitats. Less suitable habitats should be chosen only when (1) expected reproductive success in the best habitats has been sufficiently lowered by increasing density to make poorer but uncrowded habitats equally good, or if (2) resistance by already established individuals prevents entry or raises entry costs to unacceptable levels in the best habitats.

Among polygynous species, females should settle on territories where there are already females, if unmated males are still available, only when there is sufficient difference in the overall quality of the territories that their expected success is higher from bigamous matings than from available monogamous ones (Orians, 1969). A model based on these assumptions (Figure 3.2) generated a number of successful predictions about patterns of mating systems among birds and mammals. It can account for most of the available data on blackbirds (Orians, 1972), but my tests were indirect and the processes assumed by the model were not investigated. In particular, if the model is correct, the first female on a territory should have a higher reproductive success than the second, the second a success higher than the third, *et seq*. In addition, the success of a female of any given rank should be higher the larger the harem, that is, the success of first females should be higher in larger harems than in smaller ones, and so on. As pointed out by Wittenberger

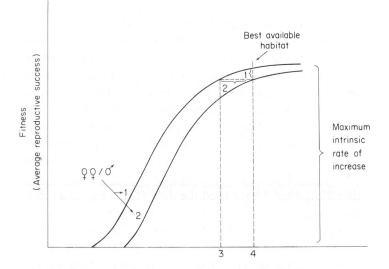

FIGURE 3.2. Model of selection of mates by females where quality of real estate held by the male influences reproductive success of females. Distance 1 is the difference in reproductive success between females mated monogamously and females mated bigamously in the same environment. Distance 2 is the *polygyny threshold*, the minimum difference in quality of habitat held by males in the same general region sufficient to favor bigamous matings by females (from Orians, 1969).

(1976) and Altmann *et al*. (1977), comparisons of average success per female as a function of harem size or rank within harems are inappropriate as tests of female choice models of mating systems.

It is difficult to tell when blackbirds have actually chosen a habitat, because adjacent uplands, which provide food for adults during early spring prior to the start of emergence of aquatic insects, are undefended. Territorial occupancy begins in early January in California (Orians, 1961) and in western Washington where Redwings are permanent residents. In areas where the species is migratory, territorial defense begins almost as soon as the first males arrive, which may be over a month before the start of breeding. Territories are at first occupied for only a short time in early morning and late afternoon. Time spent on the territory slowly increases as the season progresses until full-time occupancy is achieved. Because of the difficulty of measuring this pattern I lack sufficient data with which to test whether, within a given region, male Redwings settle first on the best marshes and only later on the poorer ones.

As it is difficult to determine when a particular female chooses a territory, the best available measure is the initiation of nesting. I have used the date of first egg as a reference point, because it can be estimated accurately for any nest that survives until hatching even if the nest was not found during the egg-laying period. At Turnbull, where there is a great deal of variation in quantity of emergence of aquatic insects, there was no correlation between date on which the first egg was laid on a lake, the median date of first eggs in all nests on a lake, and overall productivity of aquatic insects on the lake in 1965, 1966, or 1967. There are striking annual differences in the beginning of breeding but they affect all lakes equally (Figure 3.3). Also, since early nesting involves only a small proportion of females, the interval between first eggs and median date of first eggs is smaller the later breeding begins. There is also no relationship between suitability of colony site, as measured by colony size, and starting date of nests among Brewer's Blackbirds at the Columbia Refuge (Horn, 1968).

If agonistic behavior of female blackbirds during settling and nest-building phases is instrumental in deterring additional females from settling on a territory, this should be reflected in the dates of initiation of nests. Specifically, because nest construction does not normally begin until a female has been on a territory for several days or more and because construction of a nest averages about four days,

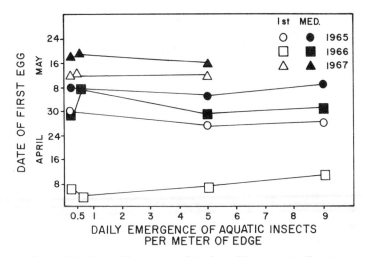

FIGURE 3.3. Date of first egg, median date of first eggs in all nests thought to represent first attempts of female Redwings, and emergence rates of aquatic insects at four Turnbull Lakes. In 1966 birds began breeding in early April on all lakes, but breeding began several weeks later in 1965 and 1967. Emergence data are for 1968, but other evidence indicates that rankings of lakes do not change much from year to year.

successive nests on a territory should be separated from one another in time by more than a week. Thus, if females are more agonistic prior to egg laying, the sequence of nesting should be more uniform than random. On the other hand, if females settled on territories without reference to the prior presence of other females, a random sequence of starting dates would be expected. For the reasons mentioned previously, this idea can be tested best by using date of first egg as the starting time for each female, even though her actual decision to settle was made at least a week earlier.

For each territory for which there are accurate data on the dates of first eggs, I calculated the mean interval between first eggs in all nests thought to represent first attempts, the average interval between dates of laying of the first egg in the first and second nest on a territory, and the average interval between dates of laying of the first egg in the second and third nests on the territory (Table 3.1). The average intervals between first eggs in successive nests on a given territory are in most cases shorter than a week in eastern Washington but longer than a week in Seattle. In many cases, intervals between the initiation of laying in first and second nests are 0-2 days. Interestingly, the interval between the first eggs in the second and third nests is usually shorter than the inverval between first eggs in the first two nests. Also, intervals tend to be shorter on territories that attracted more females than on those that attracted fewer females and shorter during years when nesting began later (Table 3.1). These data do not suggest that agonistic behavioral interactions among females are important in determining when and if additional females settle on a territory.

TABLE 3.1. Intraterritory patterns of initiation of nests. The last four columns compare mean first-egg intervals among females in harems of three or less and those of more than three females. Intervals are less than expected if females prevented others from settling while they were nest building or laying eggs.

Species / Location	Year	Avg. date of first egg on each territory	No. of days between laying of the first egg in:				Average intervals between first eggs (≤ 3 ♀♀)		Average intervals between first eggs (≥ 3 ♀♀)	
			First and second nests		Second and third nests					
			$\bar{X}$	N	$\bar{X}$	N	$\bar{X}$	N	$\bar{X}$	N
Redwing										
Potholes	1964	May 8	2.3	3	0.7	3	-	-	-	-
	1968	May 2	3.6	5	4.2	5	-	-	-	-
Turnbull	1966	April 22	8.0	40	5.7	27	9.6	18	7.9	22
	1967	May 16	3.1	39	3.4	22	3.8	24	1.7	15
Seattle	1963	April 6	19.0	4	10.3	3	-	-	-	-
	1965	April 25	8.3	4	4.0	4	-	-	-	-
Yellowhead										
Potholes	1964	May 17	6.5	16	3.5	11	7.5	10	4.8	6
	1968	May 12	4.9	9	1.6	7	6.3	4	4.0	5

Though there is no apparent correlation between productivity of a lake and time of initiation of nesting on it,
this does not exclude the expected pattern that *within* a lake
better territories get the first females. This prediction can
be tested by comparing the date of the first egg deposited
in each territory with the number of females that eventually elected to settle on the territory—an index of territory
quality, or at least an index of its attractiveness to females.

At the Columbia Refuge the number of territories is so
small that I have combined data from several lakes. For
Yellowheads in 1964 there is a correlation between the date
of first egg on a territory and the number of females settling there, with breeding beginning earlier on territories
that get more females (Figure 3.4). I have insufficient data
to test this relationship for Potholes Redwings, but at Seattle a similar relationship exists (Figure 3.5). In neither case
is the correlation very strong because many territories on a
lake attract their first females at about the same time.

At the Turnbull Refuge, Redwings bred very late in
1967, and nests were started within ten days on nearly all
territories. Therefore, any correlation between the date of
first egg and the number of females settling is difficult to
detect. In 1966, however, starting dates of first nests on
territories ranged from April 8 to May 18 and a weak
correlation does exist between the starting date and the
number of females on the territory (Figure 3.6).

Males with superior territories should be most successful
in attracting females. On the basis of emergence sampling
data, one can judge territory quality independently of size
or number of females attracted. For Yellowheads at the
Turnbull Refuge, Willson (1966) found that the number of
females attracted was inversely proportional to territory
size but directly correlated with amount of edge. At the
same location, however, for Redwings, which gather much
more of their food off territories, there is no correlation be-

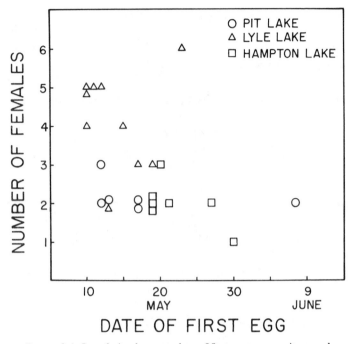

FIGURE 3.4. Correlation between date of first egg on a territory and number of females settling there. Yellowheads, Potholes, 1964. Females tend to settle earlier on territories that attract more females ($y = 5.98 - .15x$; $r^2 = .33$).

tween the productivity of a lake and the number of females nesting per territory (Table 3.2). Sex ratios among breeding birds were the same on highly productive lakes as on lakes that dried up in the summer and produced very few emerging aquatic insects. Also, the number of young fledged per female was no greater on lakes with good emergences of aquatic insects (Thirty Acre and Little McDowell) than on lakes with almost no emergence (Mann and Beaver).

Nevertheless, there are indications that the range of territory quality in an area *does* influence sex ratio among breeders. In the Redwing, polygyny is more strongly de-

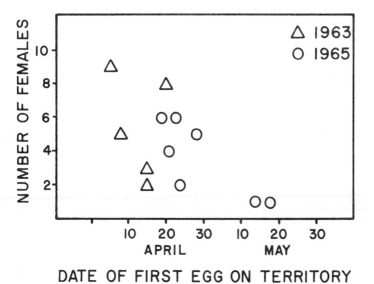

FIGURE 3.5. Correlation between date of first egg on a territory and number of females settling there. Redwings, Seattle, 1963, 1965. Females tend to settle earlier on territories that attract more females ($y = 6.60 - .13x; r^2 = .39$).

veloped in the semiarid areas of western North America, where marshes are more productive and the best territories are extremely good, than in wetter areas where marshes are less productive (Table 3.3). Also at Turnbull, where Redwings are found on lakes of low productivity and where their territories are in poorer locations because Yellowheads occupy superior peripheral territories, the sex ratio among breeders is lower than at the Potholes where lakes are more productive and there are many undefended foraging areas on lake edges where there are no emergent beds of vegetation offshore. For Yellowheads, sex ratios among breeders are also higher at the Potholes than at Turnbull (Table 3.4), probably for the same reason.

At the Potholes in 1964 and 1968, high predation rates resulted in too few nests fledging any young to permit

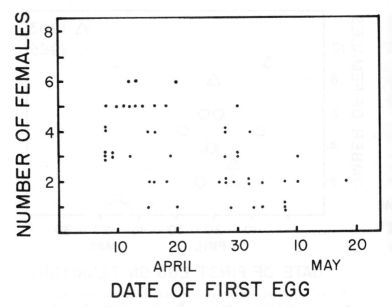

FIGURE 3.6. Correlation between date of first egg on a territory and number of females settling there. Redwings, Turnbull, 1966. Females settle somewhat earlier on territories that attract more females ($y = 4.23 - .07x; r^2 = .29$).

comparison of success of females as a function of their rank on the territory on which they bred. At Turnbull in 1966 and 1967, however, success was better on nests studied by Holm (1973). The relationship between nesting success (percentage of nests that fledged at least one young) and rank order of the female on the territory are shown in Figures 3.7 and 3.8 for 1966 and 1967, respectively. In 1966, a mild year in which Redwings started to breed early (see Figure 3.3), success was related to rank order of female and to harem size exactly as predicted by the theory of mate selection. First females did poorest on territories where only one female nested and did best on territories where three or more females nested. In addition, nesting success declined with rank over the first four

66

TABLE 3.2. Average number of females per territory and number of young fledged per female Redwing on lakes with and without large emergences of aquatic insects (data from Table 2.2). Note the lack of differences between lakes with poor and good emergences.

Lake	Emergence	1966	N	1967	N	Average
Average number of young fledged/female						
Mann and Beaver	poor	1.14	80	1.03	74	1.09
30 Acre and						
Little McDowell	good	1.23	80	0.77	65	1.00
Average number of females/territory						
Mann and Beaver	poor	3.08	26	2.74	27	2.91
30 Acre and						
Little McDowell	good	2.96	80	2.71	65	2.84

TABLE 3.3. Sex ratios among breeding Red-winged Blackbirds. Harem sizes are generally larger in subhumid or semiarid environments.

Location	No. of ♂♂	No. of ♀♀	Average ♀♀/♂	Reference
Humid Environments				
Costa Rica	16	42	2.6	Orians, 1973
Chesapeake Bay				
(tidal marshes)	126	243	1.9	Meanley and Webb, 1963
Ithaca, New York				
(uplands)	26	48	1.9	Case and Hewitt, 1963
(marshes)	16	39	2.4	Case and Hewitt, 1963
Illinois	23	37	1.6	Smith, 1943
Illinois	40	110	2.8	Smith, 1943
Madison, Wisconsin	25	49	2.0	Nero, 1956a
Western Washington				
1963	5	27	5.4	This study
1965	7	25	3.6	This study
Subhumid-Semiarid Environments				
Oklahoma	50	243	4.9	Goddard and Board, 1967
California	29	108	3.7	Orians, 1961
California	13	37	2.8	Orians, 1961
TNWR, Washington				
1966	53	160	3.0	Holm, 1973
1967	51	138	2.7	Holm, 1973
CNWR, Washington				
1964	6	35	5.8	This study
1965	5	28	5.6	This study
1966	3	14+	4.7	This study
1968	5	38	7.6	This study

TABLE 3.4. Sex ratios among breeding Yellow-headed Blackbirds.

Location	No. of ♂♂	No. of ♀♀	Average ♀♀/♂	Reference
TNWR	13	40	3.08	Willson, 1966
CNWR				
1964	29	126	4.35	This study
1965	20	80+	4.00+	This study
1968	10	41	4.10	This study
CNWR, Total	59	247+	4.09+	This study

females on the territories, though fifth-ranked females did better than fourth-ranked females in harems of 5 and 6.

In 1967, however (a cold spring), these relationships were reversed (Figure 3.8). The first females did poorest on all territories and, surprisingly, did even more poorly on territories that eventually attracted more females. These results are incompatible with processes postulated in the model of mate selection. They could be explained by postulating that the unusually cold weather caused the normally best territories to be no better than normally poor territories. Then, if settling birds assumed that the quality of territories would be normal, higher densities on the more attractive territories could have resulted in poorer success there, which affected the earliest settling birds the most. Whether or not this interpretation is correct, it is clear that patterns of reproductive success may change markedly from year to year, and many years of data are needed to determine average patterns and their causes.

Another interpretation of the contrasting results is that females begin to nest as soon as they have accumulated sufficient energy to produce a clutch. As late spring temperatures are generally not predictable from early spring temperatures, early nests can do either better or worse than subsequent ones depending on whether conditions are good or poor when the young are being fed. Also, the data represent only the success of first broods and not total seasonal production of young. A female starting earlier,

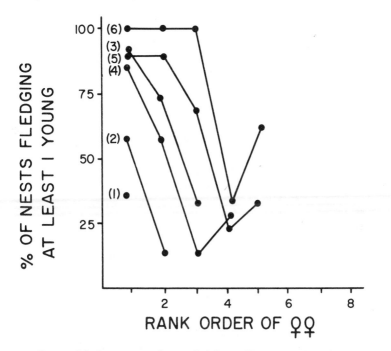

FIGURE 3.7. Percentage of nests fledging at least one young as a function of rank order of female on a territory. Redwings, Turnbull, 1966. Harem sizes are indicated by the numbers in parentheses. Success is positively correlated with both rank in harem and harem size. (Data supplied by C. R. Holm.)

even if her success with her first brood is low, has a better opportunity to breed again than a female starting later.

Finally it needs to be stressed that the relevant statistic in all of these models is the number of offspring surviving to the next breeding season and their reproductive success, not the number of young fledging. I cannot exclude the possibility that postfledging survivorship varies in some consistent manner with harem size. Because it is extremely difficult to follow survivorship of young once they leave the nests, these potentially important differences cannot as yet be assessed.

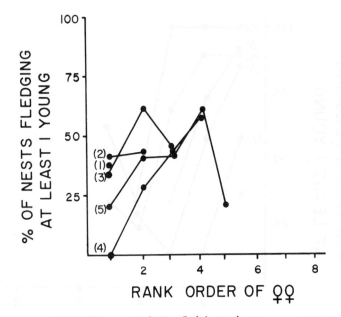

FIGURE 3.8. Percentage of nests fledging at least one young as a function of rank order of female on a territory. Redwings, Turnbull, 1967. Harem sizes are indicated by the numbers in parentheses. The pattern is the reverse of that in 1966. (Data supplied by C. R. Holm.)

Better territories could provide better foraging conditions, better nest sites, better roosting sites, or some combination of them. It is possible to assess these components by examining starvation and predation rates of nestlings on territories with different harem sizes. At Turnbull in 1966 and 1967, the percentage of nests that were victims of predators decreased with increasing harem size (Figure 3.9). This could be due to differences in nest-site quality, male antipredator behavior, or both. Studies are underway to distinguish between these two possibilities.

In contrast, there is a slight tendency for the starvation rate of nestlings to increase with increasing harem size (Figure 3.10), suggesting that females are mutually inhibit-

70

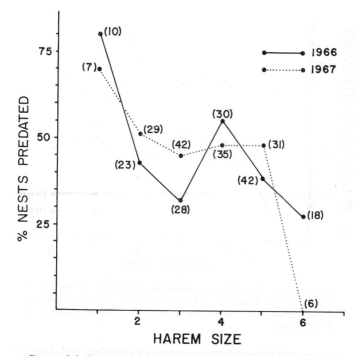

FIGURE 3.9. Percentage of nests predated as a function of harem size for Redwings at Turnbull Refuge. Sample sizes are indicated by numbers in parentheses. Predation rates are higher in smaller harems both years. (Data supplied by C. R. Holm.)

ing their ability to find food. This effect is, however, small compared to the lower predation rates and is consistent with the evidence, to be presented shortly, that there is no correlation between territory size, number of females attracted, and lake productivity at Turnbull.

Though I have examined a number of predictions which assume that females are exercising some choice in their selection of territories and males, it is important to determine the expected pattern of settling for a hypothesis of no female choice. The distribution of ♀ ♀/♂ is expected to approximate a binomial distribution around its mean value,

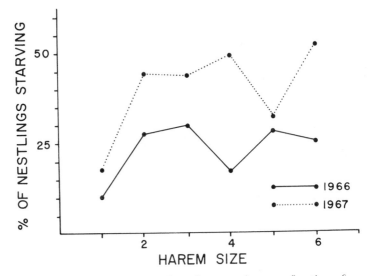

FIGURE 3.10. Percentage of nestlings starving as a function of harem size for Redwings at Turnbull Refuge. Starvation rates increase slowly with harem size in both years. (Data supplied by C. R. Holm.)

and one can thus determine whether the observed distribution is more highly clumped than the binomial. This is tested by calculating a X^2 value as follows:

$$\sum_{\text{males}} \left(\frac{\text{females attracted} - \text{average}}{\text{average}} \right)^2 \text{or } \Sigma \left(\frac{O - E}{E} \right)^2$$

with degrees of freedom one less than the number of males. Results of this test are shown in Table 3.5. Surprisingly, only the 1975 data from the Potholes differ significantly from the null hypothesis, and these numbers are strongly influenced by a single male who attracted 33 females, roughly three times the number recorded for any other territory in the years of the Washington blackbird studies.

Nevertheless, there are strong reasons for believing that

TABLE 3.5. Tests of the null hypothesis of no female choice in blackbirds by examination of the distribution of number of females attracted to individual territories (see text for details).

Species	Location	Year	No. of ♂♂	No. of ♀♀	Average ♀♀/♂	χ^2
Yellowhead	CNWR	1964	27	71	2.6	25.5 n.s.
Redwing	CNWR	1975	14	107	7.6	793 highly sig.
	TNWR	1966	53	160	3.0	41 n.s.
	TNWR	1967	51	139	2.7	34 n.s.

females *do* exercise choice in their selection of mates and real estate. First, certain territories consistently attract more females every year than other territories. Second, harem sizes are correlated with vegetation features such as amount of edge. Third, average nesting success per female does not decline with harem size as it should if females were settling randomly on territories of variable quality. Fourth, there are correlations between the starting date of first nests and the number of females attracted to a territory that are difficult to explain under a hypothesis of no female choice. I suggest that the failure of most distributions to deviate significantly from that expected if there were no female choice is that variations in territory quality are themselves approximately randomly distributed. If this were the case, a simple test of the null hypothesis by the method employed in Table 3.5 would fail to detect the important patterns revealed by the tests employed earlier in this chapter.

3.4. SIZES OF TERRITORIES

Defense of a breeding territory is a major preoccupation of male blackbirds during spring. At this time they are not under other pressures to accumulate energy, and their food supply is improving. Also, it appears that adults of the species I have studied are subject to minimal risk from predators while on their territories. I have never witnessed a successful attack on a territorial blackbird by a predator, though I have seen many attempts. The behavior of the birds also provides indirect evidence that they are not at substantial risk. Even when not actively displaying, they tend to sit in conspicuous positions with their bright patches clearly visible. Therefore, we can probably assume that the prime considerations in territorial behavior relate to the quality of the area for raising offspring and, consequently, for attracting females.

As the period during the breeding season when the rate of resource harvesting is probably most important is when nestlings are being fed, the value of food resources depends on their locations relative to the nest. More distant food sources are worth less than closer ones because the time and energy costs of traveling must be subtracted from the rate at which food is encountered and captured. The maximum size of a territory, in the absence of any external pressures, is set by the distance at which average rate of energy consumption during traveling *and* foraging times equals the harvesting rate, that is, the point at which a foraging adult utilizes all the food energy it captures simply to maintain itself. The lower value of more peripheral food sources means that they will be relinquished first to competing individuals. Moreover, the extent of pressure from other birds should in general be positively correlated with availability of resources on the area, because benefits accruing from a takeover are correspondingly higher. Conversely, when resource availability is low, the minimal area that provides enough for reproduction is larger, and outside pressure is also expected to be less.

Territories may also provide nest sites of varying quality and, depending on their size, can influence the mean distances among nests. There are differences in the kinds and numbers of nest sites provided by territories of Redwings and Yellowheads, but it is difficult to demonstrate clear relationships between the vegetational features of territories at Turnbull or the Potholes and the number of females nesting within them (Holm, 1973). Nest site features, such as depth of water, height above water, and density of cover, are known to be of importance to nesting success in populations of Redwings (Goddard and Board, 1967; Holm, 1973; Robertson, 1972). Investigations are currently underway to assess importance of nest site quality variations on blackbird breeding success at the Potholes.

If the probability of successful breeding is influenced primarily by the food resources provided by a territory, then territory sizes should be inversely correlated with food production, because the combination of increased pressures from other individuals and the capacity of productive small areas to yield high rates of prey capture should cause greater compression of territories in good locations. This tendency might be partially countered by the greater investment in aggression a male could make if his foraging opportunities were better. The emergence sampling program identified lakes that are more prolific producers of emerging aquatic insects. It also revealed that edges of patches of cattails and bulrushes have higher rates of emergence of aquatic insects than the interiors of those patches of vegetation. As Yellowheads feed extensively on their territories and are consistently exploiting primarily aquatic prey, a relationship between productivity and territory size should be most striking in this species. For Redwings, however, the nature of food resources in surrounding areas is of more importance, especially in the kinds of sites I have studied, which tend, for convenience of study, to be small marshes where most of the territories are adjacent to dry land and without edges facing open water.

From emergence data presented in Chapter Two, independent assessments can be made of the intrinsic quality of territories on specific lakes or in particular kinds of situations. These data show that there are (a) great differences among lakes in total emergence of aquatic insects, (b) that edges of lakes behind beds of emergent vegetation are poorer for emergence than edges exposed to open water and (c) that outer edges of beds of emergent vegetation are better foraging locations than interiors of beds. From this information we predict, assuming that marginal costs of territory defense increase with territory size while marginal benefits decrease, that (1) territories should be smaller on more productive lakes, (2) territories with more edge, that is interface between emergent vegetation and open water, should be smaller than territories with less edge, and (3) territories whose principal edge is the lake shore behind beds of emergent vegetation should be larger than those whose edge fronts on open water.

Predictions on a geographical scale are more difficult to make because lake productivity is inversely correlated with upland productivity and decreasing quality of a territory may be counteracted by increasing quality of nearby undefended foraging areas. The recent spread of Redwings into the uplands in eastern North America provides a good test situation, though, because birds in those habitats gather most of their food on their territories. Therefore, sizes of upland territories should reflect more closely the internal availability of prey than is the case with marsh-nesting birds.

Prediction 1 is supported by territory sizes of Yellowheads at Turnbull (Figure 3.11). I did not measure territories in the Cariboo Parklands of British Columbia, but my general impression was that Yellowheads were much more densely packed on the most productive lakes (Orians, 1966). There is no possibility of a test of this prediction at the Potholes because all lakes on which Yellowheads nest have a high rate of emergence and all territories are small (Table 3.6). The smallest Yellowhead territories recorded have been in situations where territories were providing

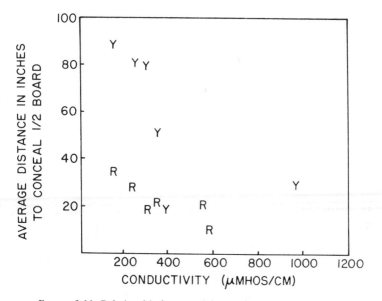

FIGURE 3.11. Relationship between lake conductivity in μmhos/cm and vegetation density on territories as estimated by distance to obscure half of a standard board. Y = Yellowheads; R = Redwings. Slopes of the two regressions are $y = 87.47 - .07x$, $r^2 = .45$ for Yellowheads, and $y = 36.58 - .04x$, $r^2 = .65$ for Redwings. Both species accept denser vegetation on more productive lakes.

primarily a nest site, and most food was being gathered in surrounding uplands. This is the situation reported by Willson (1966) for Ramer Lake, California, and I observed a similar case in the Cache Valley, Utah, in June 1965, where hundreds of Yellowheads were nesting on small cattail islands in a lake formed by a dam on the river. They did all of their foraging on adjacent irrigated cropland. I was not able to measure territories on these islands, but the density of nests was much greater than I have encountered elsewhere. Therefore, the Yellowhead, a species whose foraging is more strongly aquatic and which obtains a large fraction of its food from its territory, seems to conform to the predictions.

Redwings, on the other hand, regularly utilize a variety

TABLE 3.6. Sizes of territories of Yellowheads. The smallest territories occur where adjacent uplands provide most of the food (Ramer Lake).

Location	No. of territories	Average size (m²)	Source
Utah (marshes)	12	120	Fautin, 1941
California, Westmorland (desert marsh)	24	116	Willson, 1966
California, Ramer Lake (cattail marsh)	75	42	Willson, 1966
Washington, TNWR, 1962	10	2,210	Willson, 1966
1963	13	456	Willson, 1966
1968	9	734	This study
Washington, CNWR, 1964	27	215	This study
1968	10	217	This study

of upland habitats and upland-based prey while breeding, and regularly nest away from water in regions where uplands are productive. At Turnbull we have been able to detect no correlation between size of Redwing territories, number of females attracted, nesting success of these females, and our estimates of productivities of the lakes on which they are nesting. Redwing territories were smaller on Beaver Pond ($\bar{X} = 429$ m²; $N = 10$) than on McDowell ($\bar{X} = 1,212$ m²; $N = 8$) though Beaver has almost no emerging insects while McDowell has a very substantial emergence. Mann Lake also has no emergence, but territory sizes were similar to those on McDowell ($\bar{X} = 1,207$ m²; $N = 10$). As indicated in the following chapter, Redwings breeding at Beaver did most of their foraging on nearby Kepple Lake, and Redwings at Mann fed extensively on undefended portions of Big McDowell Lake not far away from their territories.

My previous studies in California show that under different circumstances Redwings *are* responsive to local prey availability. At the East Park Reservoir in the northern Coast Range, the lake itself was the only source of aquatic prey and surrounding uplands were mostly dry oak parkland and chaparral. Territories of Redwings nesting in isolated clumps of cattails, which were adjacent to excel-

78

lent foraging edge not occupied by other territories, were much smaller than territories of birds nesting in the main continuous study marsh. In the main marsh, territories were smaller on the edge than in the middle of the main cattail patches, but because the edge was an inner one the effect is not very pronounced (Table 3.7).

Comparative data on territory sizes of marsh-nesting and upland-nesting Redwings were gathered by Case and

TABLE 3.7. Sizes of Redwing territories. Territories are larger in humid areas, in uplands, and in large expanses of cattails (see text for details).

Location	No. of territories	Average size (m²)	Source
Wisconsin,			
Wingra Marsh	17	320	Nero, 1956
California,			
Haskell Ranch, 1959	10	1,276	Orians, 1961
Haskell Ranch, 1960	16	797	Orians, 1961
E. Park Reservoir			
isolated cattail clumps	21	234	Orians, 1961
edge of main marsh	17	789	Orians, 1961
main marsh	22	991	Orians, 1961
New York			
Airport Marsh	14	1,053	Case & Hewitt, 1963
Inlet Valley Marsh	6	1,337	Case & Hewitt, 1963
Spencer Marsh	31	405	Case & Hewitt, 1963
upland (500 ft elevation)	20	2,309	Case & Hewitt, 1963
upland (1200 ft elevation)	29	2,088	Case & Hewitt, 1963
all marshes	51	689	Case & Hewitt, 1963
all uplands	49	2,188	Case & Hewitt, 1963
Oklahoma			
farm ponds	50	350	Goddard & Board, 1967
Washington			
TNWR			
cattail, 1966	23	332	Holm, 1973
cattail, 1967	23	419	Holm, 1973
bulrush, 1966	23	603	Holm, 1973
bulrush, 1967	20	561	Holm, 1973
all types, 1968	34	842	This study
CNWR, 1968	11	513	This study
Seattle, 1963	5	2,316	This study
Seattle, 1965	7	1,740	This study
Costa Rica			
cattail marsh	7	1,101	Orians, 1973
cattail and grass marsh	9	2,361	Orians, 1973

Hewitt (1963). They found that upland territories averaged about three times as large as marsh territories, presumably reflecting lower upland productivity compared to the marshes and the fact that upland birds had to obtain most of their food from their territories while marsh-nesting birds did not.

In areas where both Redwings and Yellowheads are present, their territory sizes are comparable, but Redwings also breed in areas where prey availability and rates of recruitment are lower. In those areas, especially where territories provide a significant fraction of the food, territories tend to be larger. Marsh territories in New York were about twice the size of marsh territories at Turnbull, while the territories I measured in Costa Rica were several times larger than those in the semiarid parts of western North America but comparable to upland territories in New York (Table 3.7). There was little emergence of aquatic insects in Costa Rica, and most of what did occur took place at night. The birds there had to rely on insects using emergent vegetation, and these could not be captured at a rate as high as that possible for temperate zone birds foraging on emerging aquatic insects (Orians, 1973).

3.5. THE CUES USED BY BLACKBIRDS IN SELECTING THEIR TERRITORIES

Even if we know the factors that determine suitability of a territory and can calculate optimal choices, it does not follow that we know how organisms actually make their choices. In fact, in estimating suitability, I have employed measurements that birds cannot possibly use. Territories are selected in spring before there is any significant emergence of aquatic insects and when all lakes in the Pacific Northwest have high water levels following the winter season of heavy precipitation. In many cases aquatic vegetation is similar among lakes of extremely different insect productivity, and the same species of emergent aquatic plants occur in lakes with a wide range of water chemistry. Nevertheless, birds are able to select their breeding habitats so that females approximate an Ideal Free Distribution, and the males are also able to recognize high-quality territories. How do they accomplish this?

At first, because I had demonstrated a rough correlation between conductivity of lake water and quantity of emergence of aquatic insects, I thought birds might be able to estimate emergence on a lake by tasting the water. With appropriate receptors they could determine concentrations of dissolved salts. However, if birds are using this method, they should be fooled by lakes with high conductivities—and, hence, predicted high emergences—but which dry out late in summer and have few emerging aquatic insects during the breeding season. For example, at Turnbull, according to this hypothesis, Yellow-headed Blackbirds should settle and establish territories on Beaver and Mann lakes because they have high conductivities and suitable emergent aquatic vegetation. Nevertheless, Yellowheads seldom visit these lakes, and we have never observed any attempt to establish a territory on either of them. It is possible that older birds know that these lakes are poor and that younger males watch them and do not settle there because the older birds are not interested, but if so, I would expect to see more exploration of these lakes than seems to be the case.

Alternatively, and more likely, a foraging bird might be able to perceive insects directly in the shallow water of lake edges early in spring prior to emergence. Insects that overwintered in late larval instars should be in the surface waters, and a foraging bird might be able to make a direct assessment of probable future emergence. I have no proof that this is the case, but newly arrived birds do spend a great deal of time foraging at the air-water interface when they are selecting territories. Females, in particular, seem to ignore the vigorously displaying males and spend nearly all their time at the edge of the water where they could make this assessment.

To select a specific location within a lake, birds could use a number of direct cues. The geometry of beds of emergent vegetation and the location of edges of lakes exposed to open water are directly perceivable. From this information the probable foraging qualities of sites can be estimated. Also, birds can forage in adjacent uplands and assess the vegetation there and the prey encountered, even though the latter may be different when their young will have to be fed.

Another important aspect of emergent vegetation is density of stalks. In general, the quality of a patch of emergent vegetation should decrease with increasing density of stems, for several reasons. First, the greater the density of stems the greater the proportion of sunlight that is intercepted before it reaches the water where it is available for submerged aquatic plants. Therefore, little *in situ* production of aquatic insects is expected in denser beds of emergent vegetation. Second, the denser the bed of vegetation, the greater the probability that emerging aquatic insects, nearly all of which move in from deeper water, will emerge within a few feet of the outer edge of emergent vegetation. This may improve foraging at the outer edge of the vegetation but will make it worse elsewhere. Third, holding emergence rate constant, increasing stem density reduces the number emerging per stem. This should reduce encounter rates with prey.

Therefore, other things being equal, more open stands of cattails or bulrushes should be better foraging areas than denser ones, just as outer parts of beds of emergent vegetation are better than inner parts. We have already seen that, among Yellowheads, territory size and number of females attracted is correlated with the position of the territory in the vegetation. That there is great variation in vegetation density is immediately apparent to anyone walking around cattail and bulrush marshes, and this is confirmed by my measurements (Appendixes A and B). The means of the various lakes are much less varied, however, because, barring recent disturbance such as heavy cutting by muskrats, a mature stand of cattails reaches a reasonably constant steady-state stem density. Of greater interest is the fact that on lakes with territories of both Redwings and Yellowheads, Yellowheads in general occupy more open vegetation while Redwings are in denser vegetation, usually closer to shore (Figure 3.11). This puts Yellowheads in better foraging positions, but it is not clear whether this is because they pick territories with more open vegetation or whether it is a byproduct of their choosing territories adjacent to open water which almost always have a lower density of stems than areas in shallower water near shore.

Because blackbirds do obtain some food on adjacent uplands, the nature of that vegetation is also relevant to the quality of a territory. Redwings forage in a variety of upland vegetation types and are adept at foraging in trees and shrubs. Yellowheads, however, forage almost exclusively on the ground in the uplands (they do use sagebrush bushes at the Potholes) and rarely forage in trees. Marshes bordered by forests should therefore be poorer sites for Yellowheads because emerging aquatic insects leave marshes in the afternoon, and fly inland where they are likely to land in trees and be unavailable to Yellowheads.

While surveying lakes in the Cariboo Parklands for the presence or absence of blackbirds, I noticed that wherever patches of trees grew immediately adjacent to a lake, I could expect to find a territory of a Redwing, even on lakes that had nothing but Yellowheads elsewhere. This led me to suspect that the presence of trees next to a marsh might inhibit Yellowheads enough to result in territorial dominance by Redwings.

To test this idea I measured, with a clinometer, the angle with the horizontal subtended by the tops of adjacent upland vegetation. Measurements were taken at inner and outer edges of territories of Redwings and Yellowheads at intervals of five paces. The results are shown in Table 3.8. It is immediately apparent that Redwings are not inhibited by tall trees near their territories. In fact, they may seek out the tallest trees as song perches. In contrast, Yellowhead territories were located almost exclusively where vegetation subtended angles of less than 30° with the horizontal. Occasionally an isolated tree stuck above 30° on the inner margin of a Yellowhead territory, but this was never regular. In no case did the height of vegetation along the inner margins of a Yellowhead territory average greater than an angle of 25°. At a number of lakes there were no Yellowheads where the band of emergent vegetation was so narrow that trees on shore rose above 30°, whereas where the bed of cattails or bulrushes broadened, there would be Yellowhead territories.

At the Potholes, where there are no trees adjacent to the lakes, the basaltic cliffs have the same effect. For example, at Herman Pond, there were always two Redwing territories situated at the base of a tall cliff that rose abruptly out of the water. At other places on the lake Yellowhead territories extended to the shore. These cliffs do not create difficult foraging conditions for Yellowheads, but apparently the latter's discrimination mechanism does not differentiate among objects that loom high on the horizon. This is not surprising, because it is extremely rare that tall objects next to marshes are not trees, and presumably there has been trivial selective pressure for a more refined discrimination apparatus.

TABLE 3.8. Angle subtended by the tallest trees adjacent to blackbird territories. Yellowhead territories do not occur where trees subtend angles greater than 30°. Outer edge of vegetation refers to the interface between emergent plants and deeper open water (N = number of territories).

Location	N	Redwing range	$\bar{X}$	N	Yellowhead range	$\bar{X}$	Remarks
TNWR							
30 Acre Lake	-	-	-	15	19-32°	27.5	All taken at boundary between RW and YH territories
30 Acre Lake (south end)	-	-	-	4	20-29°	23°	At YH nests
Beaver Pond	33	10-31°	19.8°	-	-	-	Measurements made at outer edge of cattails on west side
Beaver Pond	24	34-52°	42.5°	-	-	-	Measurements made at outer edge, east side
Blackhorse (SE side)	26	25-46°	32°	7	19-26°	22°	Measurements made at outer edge of vegetation
Blackhorse	1	38°	38°	16	19-36°	24°	Outer edge of only RW territory in area
Big McDowell	-	-	-	13	18-44°	24.3°	Readings above 30° only at extreme west end of territory
Little McDowell (north)	17	29-64°	43.7°	-	-	-	Readings at outer edge of emergent vegetation
Little McDowell (south)	17	21-32°	26.8°	43	17-31°	24.7°	Measurements made at outer edge of vegetation
Mann Lake	46	19-51°	32.3°	-	-	-	Measurements made at outer edge of vegetation
British Columbia							
Sorenson	9	8-39°	27.3°	22	7-7°	7°	
Box 22	3	15-25°	21.3°	41	6-23°	7.6°	
Westwick	16	37-50°	46.2°	25	18-33°	22.1°	Measurements made at outer edge of vegetation
Rush Lake	7	28-31°	29.4°	65	3-26°	6.9°	

3.6. CONCLUSIONS

As in most natural systems, there is a great deal of unpredictable variation in important environmental factors influencing the breeding success of blackbirds. It is difficult for a bird to know in advance where a mink will hunt, where a magpie nest will be located, where patches of best emergence will be, how water levels will vary, or what the weather will be like. Nevertheless, by using cues that are correlated with average conditions, blackbirds on my study areas appear to be able to select habitats such that reproductive success of females varies little over a wide range of environmental conditions. I also have evidence that males respond in some way to potential production of aquatic insects and are sensitive to the nature of adjacent upland vegetation.

There is strong evidence that females respond primarily to real estate in their choice of breeding sites. First, their behavior at the time of habitat selection is very suggestive. A female arriving on a territory spends almost no time watching the vigorously displaying male but quickly drops down into the vegetation and moves through it, at or near the water level. It is very difficult to observe her at this time, but she seems to be foraging and otherwise assessing the territory rather than assessing the attractions of its possessor.

Second, when males are removed from territories, females continue their breeding cycle whether or not a new male takes over the territory (Orians, 1961). Why replacement males do not destroy nests at the time of takeover is not clear, but there is at present no convincing evidence that they do so.

Third, there is evidence that some sites are successful every year regardless of which particular male occupies them (Searcy, 1977). It could be argued, of course, that only the best males get those sites, but intrinsic real estate quality is probably strongly involved because correlations between harem sizes and site are just as strong between alternate years as between successive years (Searcy, 1977).

These kinds of evidence do not preclude, however, the possibility that assessment of males is a part of female decisions. The strong negative correlation between harem size and the probability that a nest was the victim of predators on Redwing territories at Turnbull in 1966 could mean that males who attract more females have in some manner signaled their effectiveness in antipredator behavior. Experiments to test this possibility are in progress.

A settling female can assess the number of females already on a territory but it is much more difficult for her to guess how many additional females are likely to nest there. My data indicate that a female can expect to exert relatively little influence on the decisions of subsequently arriving females but those females may, in fact, exert a substantial effect on her success. If it is so difficult to estimate the number of additional females likely to settle on a territory, females should not take this factor into account when picking a territory. Nevertheless, the fact that females begin to nest on most of the territories over a very short time period could mean that females are selecting poorer-quality territories early in the season because those areas are less likely to attract additional females than better-quality territories. If, on the average, additional females exert a detrimental rather than a helpful effect on earlier females, then evidence that fewer females are likely to settle may be useful. This estimation is complicated by the fact that females whose nests have been destroyed do sometimes change territories for their renesting attempts. The probability that such females will arrive at a site depends on nest predation rates elsewhere as well as on the territory itself. Presumably, estimating those rates is nearly impossible. An important focus of future studies of selection of territories by females of polygynous species must be determination of the actual effects of later females on earlier ones and measures of the extent to which females use information about those potential effects in their choices. The great variability among lakes and among years indicates that obtaining definitive information on these processes will be very difficult.

There are general theoretical reasons for expecting real estate to be an important component of mate selection by females in birds in which males establish territories before females arrive. First, real estate is assessible at the time of mate selection, and it may be easier for females to evaluate a territory than its occupant. Males may give deceitful signals which can be discovered only with considerable investment (Otte, 1975), whereas false signals about a territory are readily checked. Also, in polygynous species in which chances of attracting additional females are good, males invest less in parental care than in monogamous species where most activities are shared nearly equally by both sexes. This means that features of real estate are more important to a female than are the features of a male. The most likely exception to this conclusion is male antipredator behavior, and though this is an extremely conspicuous form of male behavior, its effectiveness is very difficult to determine. It is even more difficult to measure differences among males in their skills at preventing predation of nests within their territories.

Our evidence indicates that factors influencing territory size in Yellowheads are simpler and more clearly related to resource availability than those affecting Redwings, where simple correlations do not seem to hold. Studies are under way to assess the nature of, variations in, and effectiveness of male antipredator behavior (D'Arms, in prep.; Searcy, 1977) and to measure the importance of off-territory resources in determining territory size and number of females attracted for both species. The latter appear to be especially important early in the spring before emergences have begun from the marshes.

CHAPTER FOUR

The Adaptations:
Foraging Behavior

Finding enough food at appropriate times is one of the most serious problems confronting animals. Foraging behavior should be under strong selective pressure because it influences (a) survival during periods of food shortage; (b) rates of delivery of food to dependent offspring or rates of growth of gonads and, hence, the number of offspring that can be produced or reared; (c) hours per day which must be spent foraging and, hence, the amount of exposure to predation if risks are associated with foraging activities; and (d) amount of time available for competing uses. Some of these arguments are independent of whether or not the population is "limited" by food availability; that is, selection for foraging efficiency should occur even if food supply usually exceeds demands.

Foraging involves a sequence of decisions by an individual, concerning where to forage, how to forage (search mode), and which prey items should be pursued. If foraging effectively precludes other activities, and is no more risky than nonforaging, as is probably the case with marsh-nesting blackbirds and most other foliage-gleaning, insectivorous birds, natural selection should favor decisions which maximize energy intake per unit *foraging* (searching plus pursuing) time. However, if pursuit of prey is an exclusive activity but search for prey can be combined with territory defense, predator detection, or search for mates, or when the act of pursuit is more risky than alternative activities, natural selection might favor decisions which maximize energy intake per unit of *pursuit* time. For example, among flycatchers, *Anolis* lizards (Schoener, 1970) and trap-door spiders (Main, 1957), search involves sitting and is probably compatible with other activities, but pursuit is exclusive, more costly, and probably riskier.

It is convenient for this study to divide foraging activities into two major types: foraging for self and foraging for offspring. In the first case, whether a predator is mobile or stationary, feeding causes its gut to fill up and hunger levels to drop (Holling, 1966). In the second case, however, the predator remains hungry and may have to return to a fixed location with its prey (Central Place Foraging). I will consider both kinds of foraging activities among blackbirds but especially foraging for offspring because most of my data pertain to prey delivered to nestlings.

4.1. THEORY

I begin with an extensive treatment of optimality models of foraging, despite the existence of considerable difficulties in testing them in uncontrolled field situations, because these theories have been invaluable to me as aids in organizing my thoughts, in making decisions about which data to gather and how to analyze them. My analyses are not uniformly informative but they have shed considerable light on foraging behavior of blackbirds. Thus, theories have values extending well beyond the particular measures employed to test them, a fact insufficiently appreciated by many scientists.

The two major types of foraging decisions an organism makes are (a) where to forage and (b) which prey to take (Charnov, 1973, 1976a, b; Charnov, Orians, and Hyatt, 1976; Holling, 1959, 1965, 1966; Krebs, Ryan, and Charnov, 1974; MacArthur, 1972; MacArthur and Pianka, 1966; Pearson, 1974; Pulliam, 1974; Rapport, 1971). As succinctly summarized by MacArthur (1972), to construct a predictive theory of these decisions it is necessary to make some assumptions about the forager and its environment. Considering the environment, we must assume that there exists a variety of types of patches, each type regularly repeated so that the forager can expect to encounter the same types of prey in the same abundances in other patches of the same type. This reasonable assumption is necessary if the forager is to have any basis for comparing its current encounter rate with prey with what might be expected elsewhere.

In the formal treatment I shall use here, based on Charnov (1973, 1976a, b), these repeating units will be termed patches, defined as a piece of the environment that is internally homogeneous, i.e., probabilities of encounter with different prey types are uniform throughout it. I further assume that the prey items are encountered and handled individually, i.e., that the prey are fine grained. My data are suitable for addressing the specific questions of which patches should be visited, how much time should be spent in any patch and, while foraging, what kinds of prey should be attacked. I shall have little to say about the exact foraging modes utilized by blackbirds and how they should change with patch type or prey availability.

The species I am comparing are similar in size and general morphology so that it is likely that they have similar harvesting capabilities for the array of prey items encountered. This assumption is strongly supported by the similarities in their overall diets (Orians and Horn, 1969).

The problem of when to leave a patch takes on a special form for Central Place Foragers because the rate of capture of prey items is strongly influenced by the number of prey already being held rather than being a simple function of resource depression within the patch (Charnov, Orians, and Hyatt, 1976), the situation typical of foragers that consume their prey where captured.

Choice of Patches and the Decision To Leave a Patch

If a predator visits many patches in a foraging bout, its energy intake over a set of patches can be written:

$$E_n = \frac{\Sigma g_i \Sigma g_i(T_i)}{u + \Sigma g_i(T_i)}, \tag{4.1}$$

where T_i = time spent foraging in patch i, $g_i(T_i)$ is the energy gain for time spent in patch i and u is the average interpatch travel time. $g_i(T_i)$ may change independently of the activities of a predator or the changes may be caused by the predator, that is, a predator may exploit its prey or change the behavior of the prey so as to alter encounter rates or capture probabilities. If $g_i(T_i)$ does not change during a foraging bout, the optimal behavior for the predator is to find the best patch and remain in it, as shown by Werner (1972) and MacArthur (1972).

However, if $g_i(T_i)$ changes while the predator is foraging in the patch, then the predator should leave when the availability of prey has dropped sufficiently that its rate of intake in the patch equals the average value in all patches that should be included in its itinerary, given the current overall prey availability in the environment. This result was first derived by Charnov (1973, 1976b), who termed it the "Marginal Value Theorem." Support is provided by the laboratory experiments of Krebs, Ryan and Charnov (1974). If the predator has precise knowledge of the qualities and locations of the best available patches in the environment, it can, of course, do better by leaving the patch it is in whenever the cost of moving to a known better patch is less than the increased foraging efficiency possible there. How much detailed information a predator is likely to possess is difficult to determine or even infer, but if the predator knows only how well, on the average, it has been doing and has only average expectations, it should employ the Marginal Value Theorem.

93

This model of patch use, though of general theoretical importance, is not the most appropriate one for my purposes because when gathering food for nestlings, a bird makes a decision to terminate a bout mostly on the basis of the size of the load already in its bill. This is because the load normally reduces capture rates of additional prey much more than resource depression caused by the bird itself. The basic decisions made by a blackbird foraging for nestlings are (a) when to stop loading and return to the nest and (b) what site to pick for its next foraging trip. The first is sensitive to decreasing capture efficiencies as a function of load size, while the second probably depends on comparing the average encounter rate on the previous trip with expectations of success elsewhere in the environment.

Prey availability in patches does change in partially predictable ways during a day, and over a breeding season, and birds have available to them potential clues about patch suitability *before* making a patch choice. For example, they can presumably know that few large insects will be available on lake edges in early morning and late afternoon. They can potentially know that emergences are smaller and later on cloudy, rainy days. They can potentially know that sagebrush bushes are full of aquatic insects during the afternoon, and so forth. How much they actually know about these patterns can only be judged indirectly by observing their use of patches.

This theory of patch choice assumes that, except for differences in rates of energy capture, all patches are similar. In the real world, however, patches may differ in risk to a forager hunting there. Thus, a large finch, even though it may be a more efficient husker of seeds than a small finch, may not be able to forage safely in an open grassland and may be restricted to feeding close to the protective cover of shrubs. Patch choice models can, with difficulty, be extended to include mortality risks associated with foraging locations, but I have not attempted to do so because I have no evidence that significant differences in risk are associated with habitat patches utilized by foraging blackbirds on my study areas.

Selection of Prey Items

Consider a blackbird that has chosen a place in which to forage and now encounters food items, each of which must be pursued and handled individually. I assume that the bird recognizes prey instantaneously and makes a choice between pursuing prey or ignoring it and continuing to hunt. First, I present a model that assumes perfect knowledge on the part of the predator, an unrealistic assumption necessary to determine the *best* possible choices it can make.

A simple equation of energy intake rate given these assumptions may be derived as follows (after Charnov, 1973, 1976a). Let E be the energy taken in during a feeding period of length t, composed of T_s (time spent searching) and T_h (time spent pursuing, capturing, and consuming all prey items). The net rate of energy intake (E_n) is

$$E_n = \frac{E}{T_h + T_s}. \qquad (4.2)$$

We also assume that there are k prey types each with the following characteristics:

λ_i = number of prey i encountered in one unit of search time.

E_i = mean net energy from one item of type i.

h_i = mean handling time for an item of prey type i.

Both E_i and h_i are taken to include unsuccessful as well as successful pursuits. Equation (4.2) also assumes that energetic costs of search and handling are the same. For a full development of E_n and h_i and a stochastic treatment of them, see Charnov (1973, 1976a). From these definitions it follows directly that:

$$E = \Sigma\lambda_i \cdot E_i \cdot T_s \cdot \rho_i$$

$$T_h = \Sigma\lambda_i \cdot h_i \cdot T_s \cdot \rho_i$$

$$E_n = \frac{\Sigma\lambda_i \cdot E_i \cdot T_s \cdot \rho_i}{T_s + \Sigma\lambda_i \cdot h_i \cdot T_s \cdot \rho_i}$$

$$E_n = \frac{\Sigma\lambda_i \cdot E_i \cdot \rho_i}{1 + \Sigma\lambda_i \cdot h_i \cdot \rho_i}. \qquad (4.3)$$

Equation (4.3) is a multispecies version of the Holling Disk Equation (Holling, 1959), where ρ_i = the probability that a prey item of type i will be pursued.

If the characteristics of the prey and predator are fixed, the only thing under the predator's control is ρ_i, the probability that it will choose to pursue an individual of the ith prey type when it is encountered. If recognition is instantaneous, E_n is maximized when $\rho_i = 0$ or 1 ($i = 1, 2, 3, \ldots . k$). If the prey types are ranked by the ratio E_i/h_i, then whether or not prey type i should be eaten is (a) independent of the abundance of type i (independent of λ_i) and (b) dependent only on the abundances of prey types with rank higher than i. If we denote E_n as $E_n{}^*$ when it includes only those prey in the *optimal set*, the optimal set of prey (those to be eaten) are those of rank such that

$$\frac{E_j}{h_j} > E_n{}^*.$$

Thus, the first-ranked prey is always taken, and prey are added to and dropped out of the diet in order of rank. That is, if prey type 5 is eaten, so should 1, 2, 3, and 4. Conversely, if prey type 6 is not eaten, neither should 7, 8, 9, . . . N. If prey type 1 is common enough (λ_1 high enough), it never pays to take lower-ranked prey no matter how abundant they are. The predator adds lower-ranked prey to its diet only when better prey become scarcer. This general conclusion has been proven in various forms by Charnov (1973, 1976a), Maynard Smith (1974), MacArthur (1972), Pearson (1974), Pulliam (1974) and Schoener (1971). Previous tests of the theory are found in Charnov (1976b) and Werner and Hall (1974).

Certain practical problems arise when attempting to test this theory. For example, by definition, each individual of type i yields the same number of calories to the predator; but in any field study the investigator must make decisions about prey categories, usually on the basis of size or taxonomy. For example, prey could be grouped into age or size classes with each class considered to be a distinct type i. The fineness of subdivision of prey attempted depends upon the nature of prey variability, the precision of measurements possible or desirable, and the goals of the investigation.

A more serious difficulty is the evaluation of encounter situations with prey. Again by definition, prey of type i have a constant probability of capture, a constant pursuit time, and a constant handling time, but these values are *not* constant for species or age classes. A difficult-to-capture prey type may be surprised in an unusually vulnerable location; disease and injury may greatly influence probability of capture; encounter distances may differ and, with them, pursuit times. The result is that a prey type (as classified by size, age or taxonomic identity) that is normally rejected should be taken under those circumstances where its proximity and/or vulnerability are such that its E_i/h_i is changed sufficiently to be higher than the average ratio for the lowest-ranked prey type that is always taken. In terms of analysis of field data, this means that we cannot expect probabilities of pursuit of any prey type that *we* can recognize to be exactly 1 or 0 as predicted by the theory. Nevertheless, values should clump near those extremes if we have ranked prey properly *and* if the forager is optimizing its energy intake per unit time.

Central Place Foraging

When individuals are provisioning immobile offspring, or storing food for future use, food is not eaten where captured but it is returned to a specific central place. Two types of problems can be posed for this type of foraging. First, we can consider that an organism has a central place and ask how it should select patches, prey and its load size. Second, we can assume a set of potential foraging patches and ask where an organism should locate its central place. Here I shall be concerned primarily with the first of these problems because nest sites for marsh-nesting blackbirds are determined by the locations of patches of emergent vegetation. For the upland-nesting Brewer's Blackbird, however, the second problem is pertinent and has been treated by Horn (1968).

Central Place Foraging (CPF) is governed by many of the same factors as noncentral place foraging. In fact, noncentral place foraging is a special case of CPF in which traveling time is zero. The basic unit in CPF is the *round trip*, consisting of an outbound trip, a foraging period and a return trip. Energy is acquired only during the second part of a round trip but is expended in all three phases. Because of load cost, a return trip is more expensive per unit distance traveled than an outbound one but, in the case of blackbirds, the weight of the load is relatively minor.

Assuming that the predator can search simultaneously for many different types of prey items but handles each item individually, and that expected time to encounter the next item of type i is not dependent on the length of time the predator has been searching in a patch, rules for maximizing the rate of delivery of energy to the central place can be formulated (Orians and Pearson, 1979).

Because travel costs are unavoidable, it is normally advantageous for a predator to return with more than one prey, but returning with a single prey item may be favored if capture of the first item sufficiently encumbers the predator's prey-capturing ability that making additional captures becomes very difficult. In general, however, holding previously captured prey items should at least slightly adversely affect a predator's ability to capture additional items. If some prey types affect subsequent captures more than other types, a predator might increase its overall rate of energy capture during a trip by initially selecting prey with minimal adverse effects on subsequent captures and only taking other prey types at the end of a foraging bout.

A graphical solution to the problem of prey selection by a single-prey loader is given in Figure 4.1. Traveling time is plotted to the left and time in the patch to the right of the

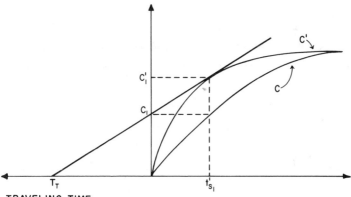

TRAVELING TIME   TIME IN PATCH

FIGURE 4.1. Prey selection by single-prey loaders. T_T = traveling time. t_s = searching time. Predator should take any prey with energy $\geq C_1$. After many trips the predator can expect an average energy = C_1'. Rate of delivery of energy to the central place is given by the slope of the line beginning to the left of the origin at a distance equal to the round trip traveling time (from Orians and Pearson, 1979).

99

origin. Time spent pursuing and capturing prey is presumed negligible in comparison to searching time. The lower of the two curves plots the energy (C) of a given prey type versus expected time (t_s') to encounter any prey possessing energy greater than or equal to C. The exact shape of the curve depends on prey densities but it must be monotone increasing if a finite number of prey types are present in the patch. The upper curve plots expected energy (C') versus expected time to encounter any prey of energy greater than or equal to C. The expected energy, C', if the predator takes any prey with energy $\geq C$ takes on a value between C and the largest prey type available in the patch and the two curves bear the relationship to one another that a tangent through any point on the upper curve intersects the ordinate at C_1, the corresponding value of the lower curve at the same point, as shown in Figure 4.1 (see Orians and Pearson, 1979, for a detailed proof). From Figure 4.1, it is evident that a predator which must spend T_T units of time traveling to and from a patch should pursue only prey of energy C_1 or greater if it wishes to maximize its expected delivery rate to the central place. As distance from patch to central place increases, the greater must be the size of prey selected by the predator, as demonstrated in Figure 4.2, in which patch quality is held constant while distance from patch to central place is varied. It is evident that if traveling time is reduced to zero, a predator should take any prey encountered, exactly as predicted for noncentral place foragers when prey-handling time is negligible. The consequences of assuming significant handling times are explored by Orians and Pearson (1979) but will not be explored here because handling times are very short for blackbirds and their typical prey.

A solution to the problem of optimal load for multiple-prey loaders is given in Figure 4.3, in which the curve (C') represents the expected energy obtained during a foraging bout versus time in the patch. The curve is assumed to flat-

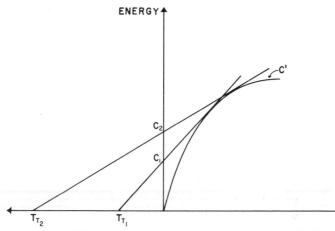

FIGURE 4.2. Optimal prey selection for single-prey loaders as a function of distance between the central place and foraging patch. If the round trip traveling time is T_{T_1}, prey with energy $\geq C_1$ should be taken. A round trip T_{T_2}, dictates a selection of prey $\geq C_2$ (from Orians and Pearson, 1979).

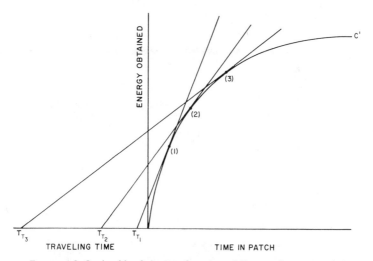

FIGURE 4.3. Optimal load size as a function of distance from central place for multiple-prey loaders. Optimal load increases from (1) to (2) to (3) as round-trip distance from central place increases from T_{T_1} to T_{T_2} to T_{T_3} (from Orians and Pearson, 1979).

101

ten slowly because there is a small loss in efficiency of capture of prey caused by the prey already held in the prey-
retaining apparatus. As shown in Figure 4.4, optimal load
increases slightly with patch quality while time spent in the
patch decreases.

4.2. TESTS OF FORAGING THEORY

I am able to test predictions from optimal foraging
theory with two different types of data. First, I have direct
observations of foraging areas and, in some cases, capture
rates of prey. Second, we have collected 1,519 food samples from nestling Redwings and Yellowheads using the
neck collar technique. From these samples inferences can
be drawn about foraging areas of the adults (Orians, 1966;
Orians and Horn, 1969). When combined with data on
prey abundances in different habitats, they provide information on preferences among prey items. Finally, from the
literature there are data on foraging activities of blackbirds
elsewhere. These data are useful in determining seasonal

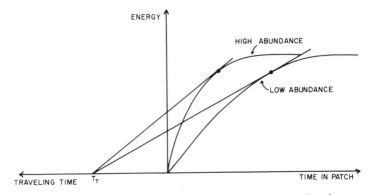

FIGURE 4.4. Optimal load as a function of patch quality for
multiple-prey loaders. Optimal load increases while optimal time in
patch decreases with patch quality (from Orians and Pearson,
1979).

102

and broad geographical patterns of foraging which supplement the more detailed data gathered on my major study areas.

Selection of Foraging Patch

Geographical Patterns. Over most of their range Redwinged Blackbirds are strongly associated with marshes, but in parts of eastern North America they now nest regularly in pastures, brushy fields, and croplands (Case and Hewitt, 1963; Robertson, 1972). These breeding birds forage primarily in uplands near their nest. In 1964, 14 samples from upland nestling Redwings in Alabama were collected by Dr. Dan Holliman on an abandoned hay field. There was a small marsh in the southwest corner of the field, but the birds foraged mostly in grassy areas close to their nests. Water levels dropped rapidly during late May and June, and the marsh was dry by the time the last young fledged. The young Redwings were fed almost exclusively on terrestrial prey (Table 4.1). Only 7 percent of the prey were from aquatic taxa (4 of 59 prey), and these were not

TABLE 4.1. Prey delivered to nestling Redwings, Robert's Field, Alabama, 1964. No pupae or tenerals were found in prey samples.

Order	Family	Larvae	Adults	Total	Order total
Collembola			1	1	1
Odonata[a]	Libellulidae		3	3	3
Orthoptera	Tettigoniidae		2	2	7
	Acrididae	3	1	4	
	?	1		1	
Homoptera	Cercopidae	8	10	18	27
	Cicadellidae	1	2	3	
	?	1	5	6	
Lepidoptera	Geometridae	2		2	19
	Noctuidae	10		10	
	?	1	6	7	
Araneida			1	1	1
Gastropoda[a]			1	1	1
	Total	27	32	59	

[a] Aquatic prey (6.8% of total).

103

necessarily gathered at the water as all odonates were adults.

Since in more humid climates uplands are relatively more productive, while marshes are relatively less productive, blackbirds should forage more away from water and bring fewer aquatic insects to their young in these regions. At Lake Wingra Marsh in Madison, Wisconsin, Snelling (1968) also found that nestling Redwings foraged primarily away from the marshes and brought mostly nonaquatic prey to their young (Table 4.2). In northwestern Iowa, Redwings delivered 53.5 percent aquatic prey and Yellowheads delivered 50.9 percent aquatic prey to their nestlings in an environment where productive hay fields were located near the marshes (Voigts, 1973).

TABLE 4.2. Food delivered to nestling Redwings at Wingra Marsh, Madison, Wisconsin (from Snelling, 1968). Data represent 113 food samples from Redwings (501 nestling hours). More than three-quarters of prey volume is derived from terrestrial taxa.

	Percent of volume of food delivered [a]			
	20 May to 10 June		11 June to 30 July	
Taxon	Aquatic	Terrestrial	Aquatic	Terrestrial
Nematoda		.03		
Annelida	.03			
Araneida	.82	2.08	1.04	1.65
Phalangida		.21		1.66
Isopoda		.53		.32
Odonata-Anisoptera	11.57		1.39	
Odonata-Zygoptera	6.03		8.71	
Orthoptera				43.21
Coleoptera	.53	.53	1.85	4.22
Neuroptera		1.69		
Lepidoptera		68.69		22.61
Diptera	4.95	.66	8.92	.44
Hymenoptera		.69		.15
Homoptera		.77		1.44
Hemiptera		.11		.27
Plant material		.11		2.51
Total	23.93	76.10	21.91	78.48

[a] Gastropods not included.

In humid western Washington, uplands are covered by dense coniferous forests, poor environments for Redwings and one in which they virtually never forage. At Foster's Island marsh in Seattle, nestling Redwings were fed primarily on aquatic insects (Table 4.3) even though aquatic prey were not abundant. Prey with aquatic stages at some time in their life cycle comprised 76 percent of items delivered to the young, and if spiders, most of which were probably taken from aquatic vegetation and water surface in the marsh, are also considered aquatic prey, then 90 percent of prey delivered to the young were marsh-associated.

The Costa Rican marshes that I studied produced few diurnally emerging aquatic insects, but the marsh vegetation supported large populations of orthopterans and lepidopterans, the principal prey delivered to nestlings (Orians, 1973). Limited evidence on abundances of insects in tropical uplands (Schoener and Janzen, 1968) indicates that prey were probably more abundant in the marsh and its edge, where blackbirds concentrated their foraging, than in upland pastures and second growth.

Turnbull National Wildlife Refuge. Upland environments at Turnbull National Wildlife Refuge in eastern Washington, consisting of wet meadows, dry, rocky grassland, pine woods, and aspen groves, were more varied than at other study sites. Insect distributions were correspondingly complex, and I am unable to assign most prey unambiguously to a particular habitat type. Therefore, I cannot analyze probable foraging areas for blackbirds from the food samples. Nevertheless, data on direct foraging observations and taxonomic composition of prey can be used to infer the importance of aquatic insects in all environmental patches (Table 4.4).

Dependence of birds on marshes for foraging can be assessed by determining the percentage of diet composed of prey with an aquatic larval stage. Not all these prey are cap-

105

TABLE 4.3. Prey delivered to nestling Redwings, Seattle, 1963-1965.

Order	Family	Larvae	Pupae	Tenerals	Adults	Total	Order total
Collembola					6	6	6
Odonata	(Zygoptera)	1		7	10	18[a]	20[a]
	(Anisoptera)				2	2[a]	
Neuroptera	Chrysopidae				1	1	1
Homoptera	Cercopidae				1	1	18
	Fulgoridae	2				2	
	Aphidae	15				15	
Heteroptera	Gerridae	1				1[a]	4[a]
	Nabidae				1	1[a]	
	Lygaeidae				1	1[a]	
	?	1				1[a]	
Plecoptera					1	1	1[a]
Coleoptera	Chrysomelidae				629	629[a]	656
	Gyrinidae	1				1[a]	
	Staphylinidae	2				2	
	Helodidae				1	1	
	Carabidae				2	2	
	?	5			16	21	
Diptera	Chironomidae			1	303	304[a]	349
	Ceratopogonidae				1	1[a]	
	Tipulidae		13		5	18[a]	
	Tabanidae	2	2			4[a]	
	Syrphidae	3	2			5	
	?	1	11		5	17	
Lepidoptera	Noctuidae	14				14	53
	Tortricidae	1				1	
	Gelechiidae	2				2	
	Pyralidae	5				5	
	?		31			31	
Hymenoptera	Braconidae				1	1	8
	Tenthredinidae	1				1	
	?	3			3	6	
Unid. Insect		2	1			3	3
Isopoda					22	22	22
Araneida					181	181	181
Chilopoda					1	1	1
Pulmonata					29	29[a]	29[a]
(Annelida)					1	1	1
	Total	62	60	8	1,223	1,353	

[a] = aquatic prey. 76 percent of prey items are from taxa with an aquatic stage in the life cycle.

106

TABLE 4.4. Direct observation of habitat types from which prey were taken by female Redwings foraging for their nestlings, Turnbull National Wildlife Refuge.

Lake	Year	% of trips to		
		Marsh	Sedge	Upland grass
30 Acre	1961	90.9	1.8	7.3
	1962	79.9	16.0	4.1
Lower Turnbull	1961	11.4	83.7	4.9
	1962	9.8	79.6	10.6

tured at or near water, but their dietary frequency shows that quality of upland foraging is strongly influenced by aquatic food production. The percentage of inferred aquatic prey delivered to nestling Redwings varied considerably among marshes and years, with an average for all lakes of 69.9 percent (Table 4.5). Yellowheads were much less variable, 91.5 percent of prey being from aquatic taxa; and at least 82 percent of nestling food items were aquatic at all lakes (Table 4.6). Unfortunately, prey sampling is insufficient to interpret causes of yearly differences, but differences among lakes and between prey species delivered, supported by some direct observations (Table 4.4), show that the percentage of upland foraging is highly variable in Redwings but follows a regular pattern for Yellowheads in most areas (Willson, 1966).

Surprisingly, in neither species is the percentage of aquatic prey delivered to nestlings correlated with lake conductivity, a rough measure of productivity (Figure 4.5). The pattern is complicated because nearby lakes influence foraging behavior of Redwings. For example, Beaver Pond is a poor producer of odonates, primarily because it nearly or completely dries up in autumn in most years, but it is very close to Kepple Lake. The section of Kepple closest to Beaver lacks emergent vegetation and, hence, lacks Yellowhead territories, making it an ideal foraging area for Redwings. Our direct observations indicate that most prey

TABLE 4.5. Percentage of aquatic prey delivered to nestling Redwings, Turnbull National Wildlife Refuge (A = aquatic; T = terrestrial). Note the great variability among lakes.

Lake	1961 A	1961 T	1962 A	1962 T	1963 A	1963 T	1964 A	1964 T	1965 A	1965 T	1968 A	1968 T	Average % aquatic[a]
Mann			.333	.667	.412	.588	.190	.810	.568	.432	.750	.250	.451
Beaver							.901	.099	.763	.237	.764	.236	.809
Kepple[b]									.602	.398			.602
30 Acre[b]	.636	.364	.200	.800	.652	.348							.496
Lower Turnbull[b]	.556	.444	.617	.383	.200	.800							.458
Reeves[b]			.364	.636									.364
Isaacson[b]			.577	.423	.800	.200							.689
McDowell[b]			.844	.156	.955	.045					.756	.244	.852
Headquarters			.982	.018									.982
Winslow			.922	.078	1.000								.961
Tritt					.838	.162							.838
Pine[b]					.850	.150							.850
Blackhorse[b]			.911	.089	.581	.419							.746
													.699

[a] The average is unweighted because differences in the number of samples reflect primarily effort expended and not population density.
[b] Lakes with breeding Yellowheads.

TABLE 4.6. Percentage of aquatic prey delivered to nestling Yellowheads, Turnbull National Wildlife Refuge (A = aquatic; T = terrestrial). Note that over 80% of prey at all lakes are aquatic in origin.

Lake	1961		1962		1963		1964		1965		1968		Average % Aquatic[a]
	A	T	A	T	A	T	A	T	A	T	A	T	
Lower Turnbull	1.000		.963	.037	.954	.046	.875	.125	.934	.066	.977	.023	.945
McDowell			1.000				.875	.125			.991	.009	.938
Kepple	.090	.091	.923	.077			.997	.003					.955
Isaacson			.852	.148	.875	.125	.733	.267					.820
Isaacson #2					.950	.050							.950
Blackhorse			.989	.011	.866	.134					.972	.028	.942
30 Acre	.928	.072	.984	.016	.878	.122	.643	.357					.858
													.915

[a] The average is unweighted because differences in the numbers of samples reflect effort expended and not blackbird population density.

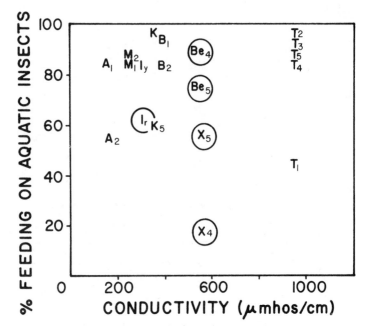

FIGURE 4.5. Relationship between lake conductivity in *μmhos*/cm and percentage aquatic prey delivered to nestling blackbirds at Turnbull National Wildlife Refuge. Values for Redwings are circled, those for Yellowheads are not. Subscripts refer to years. A = 30 Acre Meadow; M = McDowell; I = Isaacson; K = Kepple; B = Blackhorse; Be = Beaver; X = Mann; T = Lower Turnbull. Note the extensive annual variations in percentage aquatic prey delivered.

delivered to nestlings at Beaver Pond were gathered on Kepple Lake, and the percentage of aquatic prey is higher at Beaver than at any major Redwing lake at Turnbull.

The failure of Redwings to take more aquatic prey on more productive lakes is also related to the exclusion of Redwings from better foraging areas by Yellowheads. Though lakes with Yellowheads are better producers of aquatic insects, Redwings on lakes with Yellowheads (Kepple, Thirty Acre, Lower Turnbull, Reeves, Isaacson, McDowell, Blackhorse and Pine) brought only 63.2 percent aquatic prey to their young while Redwings on lakes without Yellowheads (Mann, Beaver, Headquarters, Winslow) brought 80.3 percent aquatic prey to their young.

At all lakes Yellowheads brought primarily aquatic prey throughout the day. Though birds did forage regularly in upland grass, especially in the early morning and late afternoon, these were times of maximum availability of aquatic insects in those places, and birds continued to capture primarily aquatic prey. Among Redwings, extreme patterns are illustrated by Beaver Pond and Mann Lake, both poor producers of aquatic insects but differing in availability of nearby lakes providing good foraging opportunities. At Beaver Pond, the diurnal pattern strongly reflects the emergence cycle of odonates while at Mann Lake, where foraging is primarily in the uplands most of the day, there are minimal diurnal shifts in prey taken (Figure 4.6).

Columbia National Wildlife Refuge. In the Potholes, where marshes and uplands are simpler, analysis of foraging areas can be more precise. Edges of marshes and emergent vegetation should be preferred foraging areas during the emergence period. Best foraging sites at other times are harder to determine since it is impossible to compare prey densities and probable encounter rates from our sampling procedures. However, it is clear that food availability at the edge of the water is lowest in early morning. Uplands receive their major input of aquatic insects during the afternoon, but availability of prey may also be high in the early morning when insects, many of them still not fully hardened adults, are inactive and relatively vulnerable to predation. In addition, because of rapidly changing patterns of food distributions, blackbirds may find it profitable to engage in a significant amount of exploratory foraging in all habitats.

Intensity of utilization of marsh edges was measured by direct observation at Coot Lake in 1969 from a low bluff from which all the foraging blackbirds at the water's edge

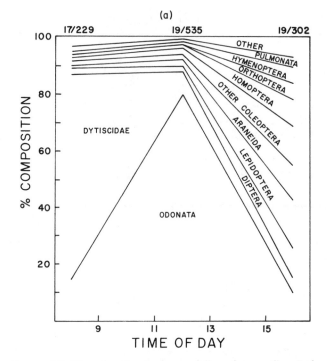

FIGURE 4.6. Diurnal patterns of prey delivered to nestling Red-wings, Turnbull National Wildlife Refuge. (a) Beaver Pond, 1965; (b) (facing page) Mann Lake, 1964. Values plotted are percentage of prey items found in the different prey categories. Number of prey items (larger number) and number of food samples (smaller number) are given above each sampling period.

could be seen and counted. Counts were made at 2-hourly intervals on days when emergence traps were visited to measure the diurnal pattern of insect emergence. All three species of blackbirds peaked in abundance at noon (Figure 4.7). Details of the pattern on Coot Lake are influenced by the fact that no birds were actually breeding there, and all came in from breeding sites elsewhere. There was a large nesting colony of Brewer's only 300 meters from the lake, but there were relatively few Redwings breeding on adja-

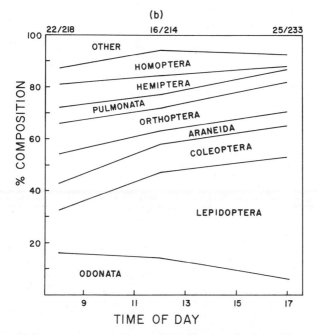

cent lakes. A population of Yellowheads breeding on Herman Pond, about 1,100 meters south-southeast of Coot Lake, was the source of virtually all Yellowheads. Probably Yellowheads foraged at edges of water closer to their nests all day since food samples taken at Herman contained primarily aquatic prey throughout the day. The willingness of birds to fly the considerable distance to Coot Lake is doubtless due to the fact that Herman Pond was carp-infested in 1969, and foraging opportunities closer to the nests were poor. In 1968 Yellowheads were not observed to fly as far as Coot Lake from active nests at Herman Pond.

Prey delivered to nestlings provides additional evidence of foraging-patch locations of adults at different times of day (Orians and Horn, 1969). Yellowheads delivered primarily aquatic prey throughout the day, but both Redwings and Brewer's concentrated on aquatic insects during

113

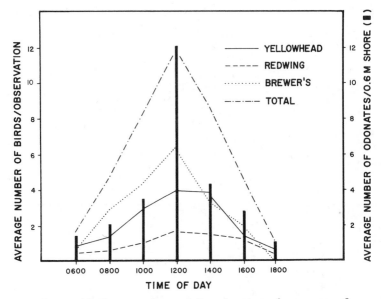

FIGURE 4.7. Correspondence of diurnal pattern of emergence of odonates, primarily damselflies, and number of foraging blackbirds at Coot Lake, Potholes, 1969. Average numbers of odonates emerging per 0.6 m of shore are plotted as vertical bars. Foraging birds were counted at the same time traps were checked to remove emerging insects.

peak emergence periods of odonates and brought mostly terrestrial insects during early morning and afternoon periods (Figure 4.8).

Using the assignment of prey to different patches employed by Orians and Horn (1969), I have calculated the diversity of patches in which birds were apparently foraging during the day (Table 4.7A). The unexpected result is that the Yellowhead, which was least diverse in kinds of prey brought, was nonetheless most diverse in the variety of patches exploited. None of the species show the expected marked reduction in diversity of patches exploited during the emergence period when lake edges have dense and rapidly renewed prey populations. There are several

114

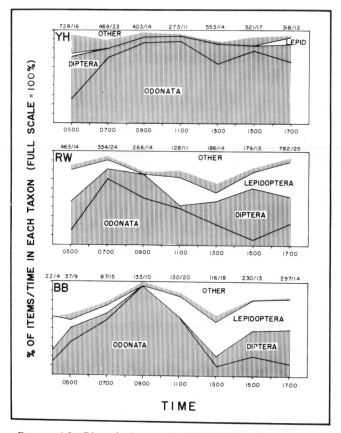

FIGURE 4.8. Diurnal changes in food delivered to nestling blackbirds at the Potholes during 1963, 1964 and 1968. Number of prey items (large number) and food samples (smaller number) are indicated above the diets for each species. The aquatic insect portions of the major prey categories are hatched to indicate the general importance of aquatic feeding for all three species of blackbirds.

TABLE 4.7. Diversity of patches foraged in by blackbirds gathering food for nestlings, Columbia National Wildlife Refuge, 1964-1965. $(- \Sigma p_i \log_{10} p_i)$

	Patch diversity of foraging during (hours)		
	0400-0659	0700-1259	1300-1900
A. Using four patch types (emergent vegetation, grass-sedge edge, sagebrush, upland dry grass):			
Yellowhead	.40	.34	.40
Redwing	.21	.35	.24
Brewer's	.31	.24	.27
B. Using three patch types (combining emergent vegetation and grass-sedge edges):			
Yellowhead	.31	.17	.31
Redwing	.15	.25	.21
Brewer's	.32	.23	.26

possible interpretations of these results. One is that we have assigned patches of the prey incorrectly. This possibility cannot be excluded as there are unproven assumptions involved in the assignment, though they are the best we can make in the light of current knowledge.

A more probable reason is that during peak emergence hours, both emergent vegetation and sedge edges of lakes are rich foraging areas. Since they constitute half of the patch types we have recognized, birds can and do forage in both, thus yielding high diversity values, except for Brewer's, which do not use emergent vegetation and whose aquatic prey are assigned strictly to marsh edges. Recalculating diversity of patch use by combining emergent vegetation with grass-sedge edge substantially reduces Yellowhead and Redwing diversity values, especially during the emergence period (Table 4.7B). Nevertheless, Yellowheads remain more diverse in their patch use except during the emergence period when they forage almost exclusively at water.

Finally, the relatively high diversity of foraging habitats utilized at all times of day may reflect the value to the birds of continuously sampling all habitats. Data have been pooled over many days and several seasons, including days in which changes in prey distribution occurred at different times. This obscures what might have been more precise patch shifts within individual days. A very intensive sampling over a short time interval during relatively constant weather conditions would help to determine the degree to which pooling of data conceals real differences in the birds' foraging activities.

We predict that birds should shift foraging areas in response to a lowering rate of prey capture in their foraging patch. At the Potholes it is easy to observe birds foraging at lake edges, and capture rates can be monitored with binoculars or telescope. In Table 4.8 I compare capture rates at the edge of Pit Lake in 1970 during emergence and postemergence periods. Though many birds shift at least part of their foraging to upland sites in the afternoon, the data reveal that individuals foraging at the lake edge captured prey just as rapidly in the afternoon as during the morning. The average size of prey may be smaller later in the day, but those observations do not discriminate between types of prey captured. However, the food samples show that more small dipterans are taken later in the day.

Another estimate of prey capture, rate of delivery of calories to nestlings, is less reliable because neck collars do not retain all the prey. For what they are worth, these data also do not reveal striking diurnal changes in rate of prey

TABLE 4.8. Rates of capture of prey by blackbirds gathering food for nestlings at the sedgy edge of Pit Lake, Columbia National Wildlife Refuge, June 1970. Capture rates are as high during the postemergence as during the emergence period.

		Capture rate during					
		Emergence period			Postemergence period		
Species	Sex	Minutes	Captures	Captures/Minute	Minutes	Captures	Captures/Minute
Redwing	♀	1.08	9	8.3	7.45	100	14.6
	♂	-	-	-	10.53	196	18.7
Brewer's	♀	19.57	264	13.2	6.57	91	13.8
	♂	12.30	176	14.3	13.33	207	15.5

capture (Table 4.9) or any consistent patterns in average size of prey captured.

It is possible that there are subtle changes in capture rates not detected by the data I have gathered, and perhaps more damselflies flush when birds approach, providing an indication that the rate of emergence is falling off after noon. Alternatively, since samples suggest that most birds are also foraging in uplands during this period, they may be responding to increased rates of capture there. I was unable to measure capture rates of prey in uplands because birds are widely dispersed and often concealed by sagebrush bushes.

British Columbia. At British Columbia I have no samples of prey abundance in different habitats, but my general observations indicate that the pattern of odonate emergence is similar to that at the two major study sites in Washington. At Westwick Lake, where there is almost no emergence and, consequently, little diurnal change in prey availability, food delivered to nestling blackbirds shows no diurnal changes. At Rush Lake, however, there were marked changes in both 1963, a year of consistently warm, sunny days during the nestling period, and 1964, when observations were made primarily during a cold, wet week. Yellowheads foraged primarily at the water during the emergence period, but they did so more strongly in 1963 when emergence was better (Orians, 1966).

Amount of foraging in different habitats by a bird during the hour period of a food sample was assessed by classifying each food sample on the basis of the prey it contained, as to whether food was gathered at water only, on uplands only, or in both locations (Table 4.10). During midday, birds foraged mostly at the water, none spent as long as an hour exclusively in uplands, and most of them never tried uplands during that period. The effect was more striking in 1963, a good emergence year, than in

TABLE 4.9. Rate of delivery of food to nestling blackbirds, Columbia National Wildlife Refuge, 1963-1965, as estimated from neck collar samples. The data are least reliable for Brewer's, where loss rates of prey are evidently higher with this method than for the other two species.

	Time interval beginning at:								
Brewer's (1964-1965)	0430	0600	0800	1000	1200	1400	1530	1700	1830
Calories/hour	278	557	370	550	338	457	828	966	676
Items/hour	5	5	4	10	5	5	17	15	23
Avg. cals/item	55	116	97	54	71	99	50	63	30

	Time interval beginning at:						
	0500	0700	0900	1100	1300	1500	1700
Redwing (1964-1965)							
Calories/hour	1632	948	2610	1539	1096	1250	1293
Items/hour	30	13	46	11	9	16	28
Avg. cals/item	54	73	57	140	122	76	46
Yellowhead (1963) (Lyle Lake)							
Calories/hour	727	-	1814	2057	1150	1077	800
Items/hour	6	-	20	36	14	22	10
Avg. cals/item	121	-	91	57	82	49	80
Yellowhead (1964-1965)							
Calories/hour	1344	1185	-	1055	1876	1361	1363
Items/hour	24	18	-	19	41	22	24
Avg. cals/item	56	66	-	56	46	62	57

119

TABLE 4.10. Number of Yellowhead food samples indicating foraging in aquatic and upland patches of the environment, at different times of day, Rush Lake, British Columbia. During midday most birds apparently fed entirely at the water during the sampling period whereas earlier and later they sampled both aquatic and upland habitats.

| | Time of day | | | | | | | | | | |
| | Early | | | | Mid | | | | Late | | |
Year	No. of samples	% Aquatic only	% Upland only	% Both	No. of samples	% Aquatic only	% Upland only	% Both	No. of samples	% Aquatic only	% Upland only	% Both
1963	15	.20	.13	.67	7	.86	0	.14	13	.31	0	.69
1964	27	0	.11	.89	35	.63	0	.37	29	.14	0	.86

1964. In contrast, few birds confined their foraging to lake edges early or late in the day, and most of them sampled both habitats during every 1-hour period. Thus, when the difference in prey availability in the two areas was less, birds engaged in much more sampling of both environments. It would be valuable to have sequences of foraging activity of individual birds to assess further the frequency of shifts, but this requires an intensity of observation that was not possible during this study.

Food samples from Rush and Westwick Lakes give a strong indication of diurnal changes in prey capture rates (Table 4.11). Delivery rates were very low at Westwick Lake and were correlated with high nestling mortality due to starvation (Orians, 1966). Delivery rates were high at Rush Lake both years in early and midday periods, but they declined sharply in the afternoon. The afternoon decline is readily explained by the emergence pattern of damselflies, the major food delivered to nestlings, but high morning delivery rates are less readily understood. Three possible explanations are that: (1), insects, though less abundant in the early morning, are nonetheless easier to capture because they are colder and less active; that (2), at the high latitude of the Cariboo Parklands, nights are extremely short in mid-June when sampling was carried out, and emergence of damselflies probably begins earlier than further south in Washington where nights are longer (if so, capture rates prior to 0500 hours might be lower, but I

TABLE 4.11. Comparative rates of delivery of energy to nestling Yellowheads, British Columbia, 1963-1964, as estimated from food samples.

| Lake | Year | Calories delivered/nest/hour | | |
		Early	Mid	Late
Rush	1963	950	900	340
	1964	1,055	1,205	630
Westwick	1964	35	-	55

121

failed to sample enough nests that early); or (3), that nestling energy needs are greater in early morning after a cold night, and adults are working harder and feeding themselves less at this time.

Selection of Prey Items

To test predictions about prey selection within patches for predators that capture and consume prey items individually, data are required on encounter rates with prey of different types and prey calorific values. Representatives of all major prey were burned in an oxygen bomb calorimeter (Appendix C). I used these data as relative measures of expected energy gain to a blackbird from consumption of individuals of these prey types. There is a potential source of bias due to variations in the percentage of chitin as a function of insect size, but this source of error is probably small. Actual encounter rates are more difficult to determine. I use data on emergence rates as an estimate of encounter rates at the edges of ponds and in emergent vegetation. These data indicate that large insects, especially odonates, are much more abundant in those two patch types during emergence periods (0700-1300 hours) than before or after. Therefore, if lower-ranked prey are being added to or subtracted from the diet in accordance with abundance of higher-ranked prey, a comparison of prey taken during these periods should reveal general patterns of decisions.

Very large prey (dragonflies, cicadas, large dytiscid larvae) are rare and, except for dragonflies, which are available primarily at dawn, do not change much in abundance during the day. Damselflies, however, yield about 50 calories and *do* change strikingly in abundance. Therefore, I have taken 50 calories as an appropriate cutoff point for testing the theory. If prey 50 calories or larger in size are sufficiently abundant, then blackbirds should reject smaller prey, but if larger prey are less abundant, smaller prey should be included in the diet. Fortunately, at lake edges the common dipterans, the most abundant small prey, are much smaller than damselflies, averaging only 10-15 calories per individual. This substantial difference makes it easier to estimate the abundance of larger prey required to make it profitable for birds to reject smaller ones.

For two prey types, the basic equation (4.3) becomes:

$$\frac{E}{T} = \frac{\lambda_1 E_1 + \lambda_2 E_2}{1 + \lambda_1 h_1 + \lambda_2 h_2}$$

if both are eaten (i.e., $\rho_1 = \rho_2 = 1$).

As by definition

$$\frac{E}{T} = \frac{\lambda_1 E_1}{1 + \lambda_1 h_1}$$

if only prey type 1 is eaten, both prey types 1 and 2 should be eaten if

$$\frac{E_2}{h_2} > \frac{\lambda_1 E_1}{1 + \lambda_1 h_1}.$$

Assume that $h_1 = h_2 = h$, that is, pursuit and handling times are independent of prey size, which is true for blackbirds in the size range being considered here. Then, since

$$\frac{\lambda_1}{1 + \lambda_1 h_1}$$

is equal to the rate of attack on prey type 1 when only prey type 1 is taken, prey type 2 should also be taken when

$$\frac{\lambda_1}{1 + \lambda_1 h_1} < \frac{E_2}{E_1} \cdot \frac{1}{h} = \frac{10 \text{ cal}}{50 \text{ cal}} \cdot \frac{1}{h} = 0.2 h^{-1},$$

if prey come in two size categories, damselflies yielding 50 calories (E_1) and dipterans yielding 10 calories (E_2).

The capture rate of larger prey (50 calories) at which it pays for a predator to include smaller prey (10 calories) in its diet is plotted in Figure 4.9 for capture times up to 8 seconds per prey item. It is difficult to estimate the average time from a decision to pursue an insect until it is captured,

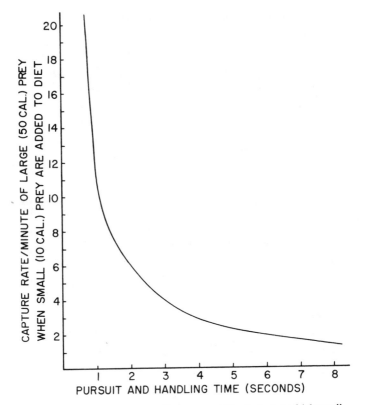

FIGURE 4.9. Capture rates of large prey (50 calories) at which small prey (10 calories) are added to or dropped from the diet as a function of average pursuit and handling times per prey item. Pursuit and handling times are assumed to be independent of prey size, which is approximately true for blackbirds foraging for insects. Above the line small prey are not taken; below the line they are.

both because the time is so short and because one cannot know when a bird first perceives a prey. Nevertheless, total time for pursuit and handling of individual prey items by foraging blackbirds probably averages no longer than 1 second and may even be shorter than that. Occasionally, however, a teneral damselfly employs escape tactics by sidling around the stem on which it emerged. I have ob-

124

served blackbirds to take up to 5 seconds to capture one such insect, but this is unusual.

Unfortunately, uncertainty about mean pursuit and handling times for prey items is of crucial importance in evaluating the theory. If handling time is 1 second, the capture rate of damselflies at which birds should add dipterans to their diet is 12 per minute. However, if pursuit and handling time is only half a second, birds should add dipterans when the capture rate of damselflies drops below 24 per minute, a rate double the previous number. Prey capture rates of adults foraging for nestlings average about 12-14 per minute, most of which are damselflies, exactly at the change point for a handling time of 1 second, making it difficult to predict optimal foraging responses. Despite these difficulties, it is of value to examine their foraging decisions, as revealed by the food delivered to the nestlings, for evidence of rejection of small prey during the emergence period of damselflies. Most of the data are presented in Appendixes D—I.

Turnbull National Wildlife Refuge. I first analyzed food samples at Turnbull by determining the percentage of prey delivered to nestlings that were 50 calories or larger in size. There is considerable interlake variation (Appendix D), but in general the percentage is higher on more productive lakes (Kepple and McDowell) than on less productive lakes, and in all cases but one, the percentage is higher during the emergence period. The percentage of samples with only prey 50 calories or larger was also greater during the emergence period (Appendix D2), but only at Kepple Lake did nearly all emergence period samples contain only large prey. These data could indicate rejection of Diptera, but they are consistent with the interpretation that birds simply took damselflies and dipterans in proportion to the frequency with which they were encountered.

The best test of the theory is to compare, within a time

period, the capture rates of large prey in samples containing both large and small prey with those in samples containing only large prey. Samples with only large prey should occur when capture rates of large prey are higher. The neck collar technique does not retain all prey delivered to nestlings, but if the loss rate of larger prey items is independent of the presence or absence of smaller items, then the comparison is valid. Contrary to the prediction, in all three time periods, capture rates of large prey are greater in samples which also contain smaller prey than in samples with only large prey (Appendix E).

For Redwings, samples from the emergence period with only large prey had slightly more prey per hour in them, but the reverse was true during the postemergence period (Appendix F). The two lakes with large numbers of samples provided different foraging opportunities for Redwings. Beaver Pond, though a poor producer of odonates, was adjacent to Kepple Lake, an excellent producer, and most birds flew to Kepple to gather food for the young. Capture rates of large prey were the same from samples with only large prey and in samples with both large and small prey (16.6 per hour and 16.3 per hour, respectively) for samples from the emergence period. In the postemergence period, capture rates of large prey were higher in samples with large and smaller prey than in samples with only large prey (9.5/hr as compared to 5.4/hr). At Mann Lake, also a poor producer of odonates, there were no alternative foraging areas nearby, and birds fed more on terrestrial insects. Capture rates of large prey were similar from samples with only large prey (6.8/hr) and from mixed samples (7.7/hr) during the emergence period, again providing no evidence that samples with only large prey resulted from rejection of smaller prey because of high encounter rates with large prey.

Columbia National Wildlife Refuge. For both Redwings

and Yellowheads in the Potholes, a much higher percentage of prey delivered during the emergence period were larger than 50 calories (Appendix G1), and a greater proportion of larger prey were delivered by Yellowheads than by Redwings. Emergence of damselflies at Lyle Lake in 1963, though not sampled quantitatively, was excellent, and large prey dominated samples at all times of day.

The percentage of samples with only prey ≥ 50 calories was high only for Yellowheads in 1963 and was generally less than 50 percent at all time periods for both species during other years (Appendix G2). Neither result suggests that birds of either species were regularly rejecting dipterans at any time of day. Rates of delivery of large prey during samples when smaller prey were also taken were as great as or greater than delivery rates of large prey when no small prey were taken (Appendixes H, I).

These data are open to several interpretations. First, blackbirds may have failed to reject small prey during the emergence period because encounter rates with damselflies were not high enough to favor such rejection. If total handling time for damselflies is actually less than 1 second, then the theoretical switch points are higher than usually observed encounter rates. Uncertainties in estimating handling time prevent further assessment of this possibility at the present time.

Second, capture rates may often be high enough to favor rejection of dipterans, but a bird may not know this at the beginning of a foraging bout. Even if a bird returns to the same general area of its last foraging bout, it is unlikely to pick the exact spot. Since emerging insects are patchily distributed, it is difficult for a bird to predict expected encounter rates at any spot. A likely policy under these circumstances would be to take all prey while making a preliminary estimate of availability of larger ones. If this were done, a bird would take some small prey at the beginning of most bouts even if it later decided to reject them.

Third, there is substantial evidence that birds are able to regulate their weights very precisely even when super-abundant food is available (King, 1972; Ward, 1965). Presumably, there are disadvantages in carrying much fat around, possibly because of poorer abilities to escape from predators and increased burdens on the circulatory system. This precise regulation implies a system of physiological feedbacks involving some measure of nutrient status. This system of regulation of feeding, though adaptive most of the year, would not be appropriate when birds were feeding young. At this time, when adults characteristically lose weight, the feedback system would signal that foraging conditions are poor. If so, lower-ranked prey would be included in the diet even if capture rates of higher-ranked prey were high. Selection should favor overriding this system of regulating food intake when young are being fed, but such a complex system of high adaptive value may not easily be completely suppressed during the short time interval when it would be appropriate to do so.

Fourth, it is necessary to consider the implication from the data that small prey items are *more* likely to appear in a food sample with a higher capture rate of large prey than in samples with lower capture rates of large prey. Data analyzed represent foods gathered by adults that do not swallow prey on capture but allow them to accumulate in the bill. If it were easier to capture small prey than large prey when a number of prey are already in the bill, or if a bird were less likely to drop prey if it added only small prey, it would pay to take small prey, if they were readily available, at the end of a foraging bout. This notion predicts that foraging bouts terminating with fewer items should be less likely to include small prey than those terminating with larger numbers of prey. Since, as will be demonstrated in the next section, more prey are taken per trip when capture rates are higher, the expected result is a positive correlation between capture rates of large prey and presence of small prey in the samples. This hypothesis could be distinguished from the second possibility mentioned above by observing in detail when small prey are taken, because hypothesis 2 predicts that they will be taken at the beginning of a foraging bout while the present one predicts that they will be taken at the end.

These two hypotheses were tested on Brewer's Blackbirds at the Potholes on May 25-26, 1975. Birds foraging for nestlings at a nest close to Coot Lake were observed with a 20× telescope. I was able to identify most (491 of 511) prey as either large (damselfly size or larger) or small (dipterans). Since foraging choices are expected to be a function of capture rates, foraging bouts were divided into those in which mean capture rate was less than 10 prey per minute and those in which it was greater than 10 prey per minute (Table 4.12).

It is clear that birds regularly took small prey at all times during individual foraging trips. For both series of foraging trips, percentage of small prey taken declined during a trip, but in the majority of bouts small prey were taken during the final five captures, and in only one of the twenty-two trips were no small prey taken at any time. Moreover, birds were as likely to take small prey during trips when capture rates were high as when they were low. On the basis of these data, I reject both hypotheses to explain inclusion of small prey in the diet at all times of the day at the Potholes.

This leaves at least two hypotheses remaining, which cannot be distinguished at present. Handling times may be slightly shorter than I have estimated, in which case optimal foraging theory predicts that small prey should always be included in the diet, as they are. Alternatively, birds may be caught in a physiological "bind" as suggested in hypothesis three. I consider an error in estimating handling times

TABLE 4.12. Time of capture of small prey during individual foraging trips by Brewer's Blackbirds gathering food for nestlings, Coot Lake, Potholes, May 25-26, 1975. Small prey are taken regularly throughout the loading process, regardless of capture rates.

No. of small prey taken during:					
First five captures		Middle captures[a]		Last five captures	
No.	%	No.	%	No.	%
A. Trips during which birds captured prey at a rate					
greater than 10 items/minute.					
1	20	1	8	0	0
5	100	3	60	4	80
1	20	12	71	3	60
3	60	6	75	4	80
3	60	0	0	0	0
13	52	22	52	11	44
B. Trips during which birds captured prey at a rate					
less than 10 items/minute.					
4	80	22	92	4	80
4	80	9	75	0	0
3	60	2	100	5	100
3	60	3	75	3	60
1	20	1	25	3	60
1	20	0	0	0	0
3	60	0	0	2	40
0	0	1	20	2	40
2	40	0	0	0	0
3	60	23	72	2	40
2	40	8	44	5	100
5	100	5	71	1	20
2	40	0	0	0	0
2	40	2	18	0	0
2	40	2	29	1	20
3	60	6	46	0	0
0	0	0	0	0	0
39	46	84	59	28	33

[a] The number of captures during the middle part of the trip is highly variable, ranging from 0 to 32. Total captures during the middle period were 42 in part A and 143 in part B.

to be more probable, but resolution of the issue must await better methods of measuring them.

4.3. TESTS OF CPF THEORY

At all temperate zone localities for which I have information, Redwings, Yellowheads and Brewer's bring more than one prey item to the nest each trip (Table 4.13). Direct observations of foraging birds reveal that they easily capture additional prey, particularly when they are foraging on soft, vulnerable emerging aquatic insects. I have not been able to measure capture rates as a function of the number of prey already in the bill because captures occur so rapidly, but many times I have seen birds with large loads drop all their prey when trying to capture another item. In such cases, birds always pick up the prey and immediately return to the nest without trying to capture additional ones.

Foraging meadowlarks bring multiple prey per trip even though many of them are obtained by digging or by gaping for them in clumps of grass. When foraging in this manner, birds put their load on the ground, gape or dig, and then pick up the prey when they are ready to move on. This is effective in the open vegetation in which they are foraging. It would, of course, be impossible for a bird foraging in bushes or trees or over water.

Redwings in Costa Rica bring only single prey items to

TABLE 4.13. Number of prey carried to nest per trip, Pit Lake, Columbia National Wildlife Refuge, 1970. All birds were foraging at distances greater than 100 meters from their nests.

Species	Sex	No. of trips	Total no. of prey items	Average no. of prey items/trip
Brewer's	♂	14	281	20.0
	♀	17	317	18.6
Redwing	♂	11	244	22.2
	♀	19	240	12.7
Yellowhead	♀	4	75	18.7

131

the nest each trip, primarily orthopterans and lepidopteran larvae that are captured either by active pursuit or by gaping them out of the vegetation near the ground or water surface. In both cases capture of additional prey would appear to be incompatible with retaining previous prey, either because of prey escape behavior or because gaping must take place over water. As expected, Costa Rican Redwings are very size selective in prey they deliver to the nest, bringing few prey less than 50 calories despite the fact that smaller individuals are much more abundant than larger ones (Figure 4.10).

At the Potholes, counting only prey actually loaded for delivery to nestlings, there is a correlation between the

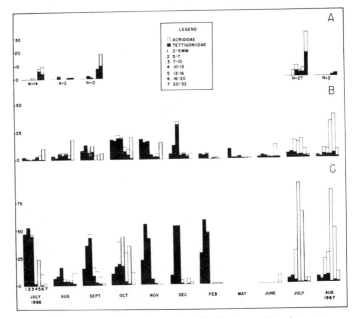

FIGURE 4.10. Comparison of size distributions of orthopterans from Redwing food samples (A) with those obtained in sweep samples in aquatic grass (B) and edge of marsh (C) at Taboga, Guanacaste Province, Costa Rica. N = the number of food samples (from Orians, 1973).

number of prey taken and the rate at which prey were being captured for male Brewer's and Redwings of both sexes but not for female Brewer's (Figure 4.11). The strongest correlation exists for a male Redwing who was feeding a fledgling at a fixed distance from his foraging area. The other graphs are based on observations of a number (unknown) of different individuals at varying (also unknown) distances from their nests. Therefore, much of the scatter in these data may be due to the varying distances from nests, because larger loads are expected from birds further from their nests and *vice versa*. Further studies with marked individuals whose nest locations are known will be required to test this hypothesis.

Another source of scatter is that I could not be certain that birds were actually returning to their nests when they left. Some may have moved to hunt in a different area. This is especially likely for those loads of less than ten insects, which were always ones when foraging had been very poor. Under such circumstances a shift to another foraging site may have been advisable. Again, this difficulty could be resolved by knowing the locations of nests of foraging birds.

A blackbird should return to its nest with a smaller load the closer to its nest it is foraging. In 1970 data were taken entirely from birds foraging at distances greater than 100 meters from their nests, and for no individual bird was the exact location of its nest known when foraging observations were made. To test the predicted relationship between distance from nest and size of load, 64 trips were observed at a nest of Brewer's Blackbirds with nestlings at Coot Lake, May 25-26, 1975. For the 31 trips in which the birds foraged within 50 meters of the nest, accurate counts of total number of prey captured during each foraging trip were made. For more distant trips it was not possible to obtain counts since the birds often foraged out of sight. How-

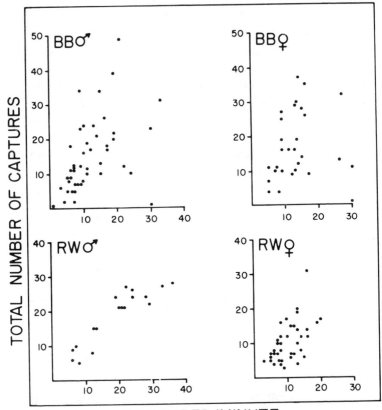

FIGURE 4.11. Relationship between total number of prey taken by a foraging blackbird before leaving a foraging area and the rate at which prey were captured during the foraging bout. Pit Lake, Potholes, June 1971. Data for male Redwings were all gathered from a single male foraging at a constant distance from his fledgling. In all other cases a number of birds at varying, and generally unknown, distances from their nests were involved.

ever, when they foraged in uplands and brought in large
orthopterans and cicadas, I was able to obtain reasonably
accurate counts as they arrived at the nest. For aquatic
loads dominated by flies and damselflies, I estimated the
load as small, medium, large, or very large, and assigned
values of 5, 10, 20 and 25 prey items respectively to them.
These values are based on extensive experience with loads
of known size.

Individuals of both sexes foraging close to the nest always returned with a small number of prey (Table 4.14). In
fact, the largest load brought by a bird foraging within 50
meters of the nest was 7 prey, despite the fact that all close
foraging trips were to the lake edge. All birds returning
from distant aquatic sites always carried large or very large
loads. For medium and distant trips, prey numbers are
presented in parentheses because I had to estimate prey
numbers in most cases. However, June 1970 data represent
accurate counts and may be taken to be representative of
load sizes for birds foraging at lake edges at great distances
from their nests.

While feeding young, adult birds must still meet their
own energy needs. Efficiency of digestion of prey might be
greatest if food were consumed regularly rather than being
ingested in a few large doses. Blackbirds with nestlings
begin most of their foraging trips by feeding themselves

TABLE 4.14. Sizes of loads delivered to nests of Brewer's Blackbirds in relation to
distance of foraging site from nest, Columbia National Wildlife Refuge. Figures
enclosed in parentheses are visual estimates and not direct counts.

Date	Sex	Close to nest (< 50 m)		Medium distance from nest (50-100 m)		Far from nest (> 100 m)	
		No. trips	prey/trip	No. trips	prey/trip	No. trips	prey/trip
May 1975	♂	6	3.4	21	(16.3)	12	(15.1)
	♀	25	2.8	—		—	
May 1976	♀	15	3.9	2	6.5	46	12.3
June 1970	♂					14	20.0
	♀					17	18.6

for several minutes before loading prey for their young. At Coot Lake in 1968, female Yellowheads fed themselves first in 77 percent of the food gathering trips (53 out of 69) while males only did so on half of their trips (7 out of 14), a significant difference ($X^2 = 4.17, p < 0.05$). Part of the difference may be caused by the fact that males feed young less frequently than females and may feed only themselves on some trips. These would be recorded as feeding by birds not provisioning young, thus giving the impression that males ate less often than females.

Interestingly, blackbirds which gathered food for nestlings used different foraging techniques than individuals which never gathered during the same foraging bout. In all cases, rates of capture of prey were higher for birds foraging for young (Table 4.15). These birds moved more rapidly, never pursued prey, and did no digging or flipping of objects. Birds that were feeding only themselves moved very slowly and spent long periods of time digging and probing in relatively small areas or short stretches of edges of lakes and streams. They probably gathered different kinds of prey, but I cannot be sure.

Since the foraging mode employed by adults gathering food for nestlings yielded higher capture rates than modes employed by birds not gathering food for young, it may be asked why the latter birds did not adopt the behavior of the former. Though I do not know the identity of prey taken by nonparental adults, my observations suggest that they were not larger than those captured by parents; therefore this advantage to the foraging mode seems unlikely.

A possible interpretation is that nonbreeding birds are under no pressures to maximize *current* rates of food capture. They can meet their metabolic needs in a small fraction of the day, and, at the time of my observations, nonbreeders were no longer attempting to take over territories or to steal copulations. Therefore, they were in a position

TABLE 4.15. Comparison of blackbird feeding rates when feeding self and gathering food for young at edge of lake, Columbia National Wildlife Refuge, 1970.

Species	Sex	Period	Feeding self			Gathering for young			
			No. minutes	No. captures	Average captures/minute	No. minutes	No. captures	Average captures/minute	Difference
Brewer's	♂	emerg.	43.75	361	8.3	12.30	176	14.3	6.0
	♂	postemerg.	46.75	428	9.2	13.33	207	15.5	6.3
	♀	emerg.	25.43	225	8.9	19.97	264	13.2	4.3
	♀	postemerg.	74.60	492	6.6	6.57	91	13.8	7.2
	♀	both	100.03	717	7.2	26.54	355	13.4	6.2
Redwing	♀	emerg.	4.65	27	5.8	1.08	9	8.3	2.5
	♀	postemerg.	12.52	131	10.5	7.45	100	14.6	4.1
	♀	both	17.17	158	9.2	8.53	109	12.8	3.6
	♂	preemerg.				0.62	5	8.1	—
	♂	emerg.				—	—	—	—
	♂	postemerg.	5.63	89	15.8	10.53	196	18.7	—
	♂	all 3				11.15	201	18.0	2.2

to use their foraging times to learn things that might be useful in the future. The foraging modes they employed were similar to those used by blackbirds under all circumstances except when young are being fed on emerging aquatic insects. Those are, presumably, the modes most efficient at revealing prey in the majority of patches and under most circumstances. Therefore, experience obtained from this deliberate gaping and probing into the substrate may have survival value during future periods of food shortage. Alternatively, adults foraging for themselves may encounter different kinds of prey that are less suitable as food for nestlings or which provide specific nutrients needed by adults, but this seems less likely.

Those birds foraging for nestlings generally ate the same kinds of prey that they subsequently loaded for their young. The major exception was dragonflies, which were never consumed by birds that loaded food for young later during the same trip. A bird foraging for nestlings always terminated its foraging trip and flew to its nest whenever it captured a dragonfly. Dragonflies contain 500 to 800 calories, depending on the species, more than ten times the value of a damselfly. Their large size may make it more difficult for a bird to capture additional prey, thereby making an immediate return to the nest advantageous. Dragonflies were regularly eaten by blackbirds not gathering food for their young, which indicates that they found them acceptable.

4.4. CONCLUSIONS

With respect to patch selection, Redwings and Brewer's shifted their patch utilization as expected in accordance with prey availability while Yellowheads did not at the Potholes but did so in British Columbia. I also obtained substantial evidence that birds were sampling different patches much more regularly early and late in the day, when prey abundances were not much greater in any one patch type, than during the emergence period when edges of marshes were substantially superior as foraging sites. Sampling different patches is most valuable when no one patch type is predictably better than others and when locations of places with high prey densities are unpredictable. However, I have been unable to determine how the birds decide to switch patches. My data on capture rates are not helpful, and, as indicated earlier, a decision to stop loading prey is not a decision to change patches. Therefore, I am unable to test theories such as The Marginal Value Theorem nor is it evident how I could do so with blackbirds.

Does the failure of Yellowheads to shift in patch and prey utilization as much as the other species, particularly at the Potholes, constitute evidence that they are not foraging optimally? There is more nestling starvation among Yellowheads (Willson, 1966; Orians, 1966) than among Brewer's (Horn, 1969; Furrer, 1974), but no more than among Redwings (Holm, 1973). Our data on rates of delivery of energy to nestlings and on nestling survival do not provide any indication that Yellowheads are less successful in foraging than are Redwings. In both species, unlike Brewer's, only a single adult delivers food to most nests so that delivery rates are legitimately compared.

The differences between Yellowheads and the other species can probably be understood only with reference to the presence of those species in the same environment. Redwings and Yellowheads are interspecifically territorial, and this influences local distribution patterns. In addition, their characteristics have evolved in relation to each other. The habitat and foraging patch selection behavior of the species would doubtless be different if one or the other had been absent. I return to this problem in Chapter Five after data on overlap in foods and foraging have been considered in greater detail.

As predicted from Central Place Foraging Theory, rejection of smaller prey was more pronounced in Costa Rica, where Redwings brought only single prey items to a nest each trip, than in the Pacific Northwest where multiple-prey items are invariably delivered. I obtained no unequivocal evidence that foraging blackbirds in Washington ever rejected small prey that they encountered, but because of the sensitivity of the switchpoint to prey handling times, and because capture rates I measured were very close to those at which rejection of small prey would be expected, these results provide no evidence for or against the foraging model. Very detailed observations of capture rates and the identity of prey captured are needed to resolve this problem.

These ambiguities in the interpretation of my data illustrate some of the difficulties in testing optimality models in the field. It was difficult or impossible for me to measure some of the necessary parameters. Also, because of confounding variables, it is difficult to judge whether or not data fit the predictions of a theory well enough to constitute some support for it. At best, all of my tests are weak because all correspondence between theory and data, even if good, could be fortuitous. Nonetheless, without the theories I would not have thought to analyze capture rates of large prey in mixed and pure samples, load sizes as a function of distance from nest, time during loading when small prey are taken, patch sampling within food samples, or size selection under single- versus multiple-prey-loading circumstances. All those analyses shed considerable light on the foraging behavior and decisions of blackbirds. Thus, the value of the theories was substantially greater than the power of the tests I was able to employ to test them.

CHAPTER FIVE

The Patterns: Variability
in Use of Resources

In the previous two chapters I dealt with attributes of individual birds that are directly molded by natural selection. In all sexually reproducing organisms, however, there are patterns of variation within and between populations that are *indirect* results of selection for other individual attributes. These patterns form the focus for the next two chapters. Some of them were part of my original study design while others simply emerged from gathering data for different purposes and were not predicted by any theories of adaptation being tested. Both types of patterns are considered together because they differ only in the kind of attention I happen to have given them.

In this chapter I examine variability in some characteristics of blackbirds—such as range of lake productivity occupied, range of nest sites occupied, and breeding seasons—that directly involve the use of a range of some environmental resource, and some—such as clutch sizes and variability in foods and foraging—that do not, but which are presumably correlated with those uses. For some traits I have measures of both the within-phenotype and between-phenotype components of variability, but for the majority I can discuss only total population variability.

5.1. RANGE OF LAKE PRODUCTIVITY OCCUPIED

Choice of breeding habitat, which is made at least once annually by each adult bird, generally commits an individual for an entire season. In addition, the decision is made early in the spring when cues about conditions when young are to be fed may be very incomplete. There is no obvious *a priori* reason why choices, especially those of females, should be constrained by those of previous seasons. Nevertheless, success at a site in one year may be a good predictor of success in subsequent years. The range of habitats occupied if individuals had complete information available to them and were able to exercise their choices freely (the Ideal Free Distribution of Fretwell and Lucas, 1969) is determined by relative fitnesses of individuals in different environments, rate of reduction of fitness due to population density, and densities of interspecific competitors that reduce fitness selectively in some habitats.

There are several reasons why reproductive success of Redwings should not drop as rapidly as that of Yellowheads along a gradient of decreasing food availability. Redwings are smaller than Yellowheads, and an individual requires less total food per day. Yellowheads, in the Pacific Northwest at least, depend primarily on emerging aquatic insects, probably because they can be captured at such high rates. Also, the birds are not effective foragers in trees. For Yellowheads, therefore, it is more difficult to breed in habitats where prey encounter rates and prey renewal rates are lower. Redwings should also do best in habitats yielding high capture rates, but they are successful over a broader range of habitats. The greater size of Yellowheads, however, gives them a competitive advantage in encounters (Orians and Willson, 1964), permitting them to exclude Redwings from their preferred areas (Miller, 1968).

In the Cariboo Parklands of British Columbia, Redwings breed on lakes ranging in water conductivity (a crude prediction of productivity) from 625 to 3,430 micromhos/cm, while Yellowheads are restricted to lakes above 1,100 micromhos/cm (Orians, 1966). Reproductive success of Yellowheads was much poorer on one lake (Westwick) close

142

to the lower conductivity limit than in one much above it (Rush), with most of the nestling losses attributable to starvation (Orians, 1966).

At the Turnbull Refuge, Yellowheads as breeders are also restricted to a fraction of the lakes. Our emergence traps captured a daily average of 16.8 odonates per meter of edge and 15.0 odonates per square meter of emergent vegetation on lakes with breeding Yellowheads (Kepple, Blackhorse, Big McDowell and Little McDowell), but only 0.4 odonates per meter of edge and 0.1 odonates per square meter of emergent vegetation in those lakes without breeding Yellowheads (Beaver and Mann) (see Table 2.2 above for details). Similarly, Yellowheads were absent from Willow (every year) and Lyle lakes (1968) in the Potholes, lakes that produced far fewer emerging insects those years than North Hampton. Yellowheads do not breed on highly productive Coot Lake because it lacks emergent vegetation capable of supporting nests. Redwings bred on all these lakes, as they do on virtually all lakes in the general region. Yellowheads are also absent from western Washington, where all lakes are low in dissolved nutrients and produce few emerging aquatic insects. Many of these lakes are too poor to support breeding Redwings.

An experiment indicating that lake productivity is the major limiting factor in western Washington was performed on Fern Lake, an oligotrophic small lake located on the Kitsap Peninsula, Kitsap County. Between August 19-31, 1965, the lake was fertilized with a complex mixture of nutrients designed to maximize primary production. There was a very rapid heavy algal bloom, a great increase in zooplankton, and great increases in the number and growth rates of chironomids in the lake sediments (Donaldson, *et al.*, 1971). During the winter the lake flushed out and most nutrients were lost, but emergence of insects, mainly chironomids, was more than twice as heavy in April and May 1966 than in previous years. Redwings had not been known to breed at Fern Lake, but on May 31, 1966, a nest with three young about one week old was found in brush at the south end (Harold Klaasen, pers. comm.). Redwings have not bred on the lake since then.

The restriction of Yellowheads to highly productive lakes is very noticeable at the eastern periphery of their breeding range where precipitation is higher and most lakes have outlets and do not fluctuate markedly or concentrate nutrients. In Wisconsin, where I know their breeding distribution fairly well, Yellowheads are restricted to a small number of lakes, known among local bird watchers as "Yellowhead" lakes. Most, but not all of them, are "prairie potholes" without external drainage. On lakes with outlets, Yellowheads breed only when some disturbance has opened up dense beds of emergent vegetation. This increases *in situ* production and emergence and offers better foraging conditions because insects are concentrated on a smaller number of stalks.

The history of Yellowheads on a marshy lake near Pardeeville, Columbia County, Wisconsin, is informative. Prior to 1958 the lake was covered by a dense stand of cattails and supported a low density of breeding Redwings but no Yellowheads. That year beavers built a dam across the outlet, raising water levels sufficiently to kill most cattails. During two years, while emergent vegetation slowly returned to its former closed status, the lake supported a thriving breeding population of Yellowheads. When I visited the lake in June 1960 it was again covered by a dense stand of cattails and had only a few Redwings. An increase in breeding populations of Yellowheads in Iowa caused by thinning of dense beds of emergent vegetation by muskrats has been documented in detail by Weller and Spatcher (1965) and Weller and Fredrickson (1974).

Variations in settling densities are proximally caused by

144

the abundance of birds relative to availability of suitable territories, and spatial and temporal variability in suitability of sites. In general, the first male blackbirds to arrive defend larger territories than they subsequently hold. Subsequent compression in response to challenges by later arriving birds occurs because the cost of defense of larger areas increases faster than benefits—peripheral food sources becoming worth less than central ones due to increasing traveling times from nests (Orians and Pearson, 1979). In polygynous blackbirds, however, spacing of females within a male's territory enhances the real value of peripheral resources and, hence, the value of larger territories.

Assuming that the genetic basis of variations in sizes of territories defended by different individuals is of similar magnitude in both species, I made the following predictions concerning territory size variability in Redwings and Yellowheads. First, there should be greater variability in Redwing territory sizes from year to year, from lake to lake, and within lakes because Redwings occupy marshes with a much wider range of abundance of emerging insects than do Yellowheads. Also, Yellowhead territories are concentrated around outer edges of vegetation, the best foraging sites. Compared to the variety of conditions encountered on Redwing territories on the same marshes and on marshes with only Redwings, Yellowhead breeding sites are certainly more uniform. Data available to test these predictions are contained in Table 5.1. Unfortunately, comparisons are restricted because I have Yellowhead data from only one Turnbull Refuge lake and Redwing data from only one Columbia Refuge lake.

Mean Yellowhead territory sizes were remarkably similar on different Potholes lakes, and territory sizes on Hampton Lake were approximately the same in 1964 and 1968. In contrast, Redwing territory sizes averaged three times as

145

TABLE 5.1. Variation in sizes of territories of Redwings and Yellowheads.

Location	No. of territories	Mean size (m²)	Range (m²)	Standard deviation	Coefficient of variation (%)
		Redwing			
Taboga, Costa Rica					
West Marsh	9	2,361	1,022-4,343	973.9	41.2
Iguana Marsh	7	1,101	487-1,932	526.1	47.8
CNWR, 1968					
Lyle Lake	6	369	112-581	164.3	44.5
TNWR, 1968					
Mann Lake	10	1,207	533-1,498	294.0	24.4
Beaver Pond	16	428	202-778	161.8	37.8
McDowell Lake	8	1,212	519-2,348	576.4	47.6
Seattle, 1963	5	2,316	1,940-3,220	522.2	22.5
Seattle, 1965	7	1,740	1,380-2,420	402.8	23.1
		Yellowhead			
CNWR, 1964					
Pit Lake	9	231	98-356	88.9	38.5
Lyle Lake	10	202	42-392	120.8	59.8
Hampton Lake	8	214	146-286	50.2	23.5
CNWR, 1968					
Hampton Lake	10	217	50-544	137.5	63.4
TNWR, 1968					
McDowell Lake	5	637	389-893	188.7	29.6

large on Mann and McDowell Lakes, where Yellowheads occupied the best sites (McDowell) or there were no nearby high-quality foraging areas (Mann), as they did on Beaver Pond, where a rich, undefended foraging area was available on adjacent Kepple Lake. However, since Redwing territories were considerably larger than Yellowhead territories at both Columbia and Turnbull refuges, coefficients of variation were similar despite much larger standard deviations in Redwing territory sizes. Thus, the data do not support the expectation of greater variation in Redwing territory sizes, possibly because the exclusion of Redwings from the best sites eliminates most potentially small territories.

Another component of settling densities in polygynous blackbirds is the number of females attracted per territory. For the same reasons given above, I predicted greater variation in number of females per male in Redwings within lakes and geographically. Relevant data are given in Tables 5.2 and 5.3. The number of females per territory ranged from 0 to 13 among Redwings and from 0 to 8 among Yellowheads. Within-marsh variations in harem sizes are not evidently different between the two species, but the mean ratios are less variable between lakes among Yellowheads, ranging from 1.7 to 4.2, whereas ratios between lakes among Redwings in Washington have ranged from 2.1 to 7.6 and were as low as 1.3 in the Iguana Marsh in Costa Rica. The mean number of females per Redwing territory at Turnbull was very similar in 1966 (3.0) and 1967 (2.7), but the amount of variation at the Potholes was greater, ranging from 4.7 in 1966 to 7.6 in 1968. The higher value

TABLE 5.2. Intramarsh variation in sex ratios of breeding blackbirds.

Location	No. of territories	Mean No. ♀♀/territory	Range	Standard deviation
	Redwing			
Taboga, Costa Rica				
West Marsh	9	3.7	0-9	2.7
Iguana Marsh	6	1.3	0-3	1.4
CNWR, 1968	5	7.6	4-13	3.36
TNWR, 1966				
30 Acre	7	3.6	1.6	1.99
Mann	17	3.4	1-6	1.54
Beaver	9	3.1	2-5	1.27
Little McDowell N.	10	3.2	1-6	1.62
Little McDowell S.	9	2.9	1-5	1.62
TNWR, 1967				
30 Acre	8	3.6	1-6	1.92
Mann	17	2.7	0-6	1.69
Beaver	10	3.3	1-6	1.64
Little McDowell N.	9	2.4	1-4	1.24
Little McDowell S.	7	2.1	1-3	0.90
Seattle, 1963	5	5.4	2-9	3.1
Seattle, 1965	7	3.6	1-6	2.2
	Yellowhead			
CNWR, 1964				
Pit	9	1.7	0-3	1.0
Lyle	10	4.2	2-6	1.2
Hampton	8	2.5	0-3	1.5
CNWR, 1968				
Hampton	10	4.1	1-8	2.1

TABLE 5.3. Annual variation in mean sex ratios of breeding blackbirds among study areas.

Location	Year	Mean No. ♀♀/territory	No. of territories
Redwing			
TNWR	1966	3.0	53
	1967	2.7	51
CNWR	1964	5.8	6
	1965	5.6	5
	1966	4.7	3
	1968	7.6	5
Seattle	1963	5.4	5
	1965	3.6	7
Yellowhead			
CNWR	1964	4.35	29
	1965	4.00+	20
	1968	4.10	10

for 1968 is based on only five territories, three of which were located in the only good remaining cattail patches in a shallow corner of Lyle Lake where carp could not enter. One of these territories had 13 females on it and, with the small sample size, this particular territory strongly influences the mean and standard deviation of harem sizes for that year. The mean for the other four territories was only 6.2.

Variations in part reflect changes in numbers of territorial males independent of changes in number of females. For example, in Seattle, the number of nesting female Redwings was similar in 1963 and 1965 (27 vs. 25), but for unknown reasons, the number of territorial males increased from 5 to 7, dropping the average females per territory from 5.4 to 3.6.

Changes in sex ratios of breeding birds were due to a decline in marsh quality at Lyle Lake from 1963 to 1968. In 1963, when I first began studies in the Potholes, Lyle Lake had the densest breeding population of Yellowheads at any Potholes lake, and there were no Redwings there. I did not make a map of the lake until after the breeding season and did not count territorial males in 1963, but in 1964 the lake supported 10 territorial males and at least 42 females (Table 5.4, Figure 5.1). A carp invasion began in 1964, increased during succeeding years, and nearly destroyed the emergence of aquatic insects (see Chapter Two). By 1965 there were only four territorial male Yellowheads and only five females nested. By 1966 there were only two territorial males and no females. In 1967 only one male held a territory, and he failed to attract a female. Concurrently with this decline in Yellowheads, some Redwings began to move into the area, but by 1968 the marsh was so poor that no nests of Redwings were built. Apparently males are more site-conservative than females, partly because they must make their decisions earlier and with less reliable information and partly because site shifts are more powerfully constrained by territoriality.

5.2. RANGE OF NEST SITES OCCUPIED

Nest sites are chosen by females, and a site once chosen is occupied until the nest is destroyed or fledges young. Renesting birds nearly always picked a new site for subsequent nests during the same season. Quality of a nest site depends on its location with respect to areas from which food for nestlings will be gathered, the protection it provides from weather and predators, and its position relative

TABLE 5.4. Changes in blackbird territories on Lyle Lake, Columbia National Wildlife Refuge, during a period of invasion of carp which began in 1964 and had nearly eliminated emergences of insects by 1968.

Year	Redwing			Yellowhead		
	$\male\male$	$\female\female$	$\female\female/\male$	$\male\male$	$\female\female$	$\female\female/\male$
1964	0	0	-	10	42	4.2
1965	1	2	2.0	4	5	1.2
1966	2	6	3.0	2	0	0.0
1968	2	0	0.0	1	0	0.0

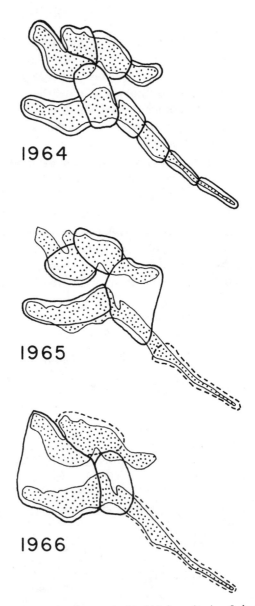

FIGURE 5.1. Yearly changes in blackbird territories, Lyle Lake, Potholes. The stippled area represents beds of cattails and bulrushes. Yellowhead territories are enclosed by solid lines, Redwing territories by dashed lines.

to other conspecific nests. Nearly all first broods of black-birds in Washington are raised in nests supported by vegetation that grew the previous summer, because new growth is usually not tall enough or strong enough to provide adequate support or cover.

In Washington blackbirds favor cattails for nests, especially early in the nesting season, because cattails are less flattened by winter rain and snow than bulrushes. They therefore provide better nesting cover and offer sites farther above the water than are available in bulrushes or still-leafless shrubs. At Turnbull, where the most complete data are available, nesting success increased with increasing height above the water (Holm, 1973), and similar results are reported by Meanley and Webb (1963) and Holcomb and Twiest (1968). In contrast, Goddard and Board (1967) in Oklahoma found that nesting success in Redwings decreased with increasing nest height, but success did increase with increasing depth of water below the nest. This may have obscured a height correlation, as cattails and bulrushes protrude further from the water in shallower than in deeper water. At Turnbull there has been no consistent pattern of nesting success in areas with cattails and those with bulrushes (Holm, 1973). Other sites have been used so infrequently that statistical comparisons are impossible.

Good foraging sites for Yellowheads are nearly always accompanied by good nesting sites in emergent vegetation, and Yellowheads seldom use other sites (Appendix J). On my study sites in Washington and British Columbia I have not found nests in anything but cattails and bulrushes, but Yellowheads also use small trees and shrubs standing in the water in some areas (Miller, 1968).

Redwings also built most of their nests in emergent aquatic plants at my study areas, but a few nests were situated in other sites (Appendix K). At the Potholes in 1965 one female actually relined a one-year-old nest of a Northern Oriole (*Icterus galbula bullocki*) and successfully raised three young in it! Variability of nest sites used by Redwings is also summarized by Miller (1968). Redwings have even been reported nesting in a bird box (Nero, 1956b) and on the ground (Glitz, *fide* Holcomb and Twiest, 1968; Horn, pers. comm.). In eastern North America, marshes have more woody plants growing in them, and these plants are often selected by breeding Redwings (Holcomb and Twiest, 1968; Weller and Spatcher, 1965).

To measure breadth of nest site utilization, I divided nest sites into three categories; trees and shrubs, coarse emergents, and forbs, and calculated a measure of width using the statistic of diversity, $H = \frac{1}{\Sigma p_i^2}$, where $p_i =$ the proportion of nests located in the ith nest site type (Table 5.5). Though Redwings construct most of their nests in emergent aquatics, they are more variable than Yellowheads in this trait. If a category of location over dry land were also added, the variability of Redwing nest sites would be further enhanced.

These differences do not necessarily reflect different rankings of sites by females, because shrubs and trees are less available within Yellowhead territories than within Redwing territories. I have no independent measures of availability of all types of nest sites, but there are many situations at both Turnbull and the Potholes where trees and shrub sites are available to Yellowheads and yet are never used. Miller (1968) suggested that the willingness of Redwings to use a variety of nest sites may be responsible for their great ecological success in a variety of environments, but I suggest that causal relations are the reverse. Redwings select a variety of nest sites because they are otherwise capable of extracting resources and breeding in a wider variety of environments than Yellowheads. This has favored the use of sites that are not good choices for Yellowheads.

Table 5.5. Diversity of nest sites utilized by Redwings and Yellowheads. (Diversity = $1/\Sigma p_i^2$.)

Study area	Source	Diversity of nest sites of	
		Redwings	Yellowheads
CNWR	this study	1.108	0.000
TNWR	this study	1.049	0.000
Cariboo Parklands	this study	0.000	0.000
Seattle	this study	0.000	-
Crab Creek	this study	0.000	-
Taboga, Costa Rica	this study	1.066	-
Iowa	Weller & Spatcher, 1965	1.156	0.000
Ohio-Michigan	Holcomb & Twiest, 1968	2.566	-
Oklahoma	Goddard & Board, 1967	1.150	-

5.3. BREEDING SEASONS

Timing and length of breeding season of any wild bird is influenced by several factors. Most birds probably breed as long as it is profitable to do so. The most important variable is probably time of maximum food supply upon which the parents feed their young (Lack, 1948, 1954a). The start of breeding in spring may be determined by ability of females to accumulate sufficient energy to build a nest and form a clutch (Perrins, 1965, 1970). Evidence that this is true for Redwings and Yellowheads was presented in Chapter Three. In Brewer's Blackbirds at the Potholes the earliest nests are most successful in fledging young, mostly because of lower rates of predation (Horn, 1968; Furrer, 1974), suggesting that birds might begin to nest earlier if they had sufficient energy reserves to do so.

Termination of the breeding season may be influenced by advantages in completing the annual molt while food resources are still favorable (Pitelka, 1958). Also, in the Potholes, brood parasitism by Brown-headed Cowbirds (*Molothrus ater*) on Brewer's Blackbirds increases substantially during the breeding season and is a major factor in reducing the success of later nests (Furrer, 1974).

Variability in timing of breeding between years should

153

be determined primarily by rate of improvement of foraging conditions in spring. An early spring should trigger earlier breeding of blackbirds on my study areas, because prebreeding foraging, which is almost entirely on the uplands, is not related to timing or quantity of subsequent insect emergences from the marshes. Because Redwings are smaller than Yellowheads, it should take them less time to accumulate sufficient reserves for breeding. This difference is striking in terms of the dates for the first nests to be started. At Turnbull the earliest known Yellowhead egg during our study was laid May 2, while one Redwing nest received its first egg on April 7. Similarly, at the Potholes, the earliest known Yellowhead egg was laid April 29, while the first known Redwing egg was laid April 6. Nevertheless, there was remarkably little difference between the two species in the mean date of the first egg in all nests believed to represent first clutches (Table 5.6). The average date of *all* first eggs in Redwing nests would be later than shown in the table if second clutches and renestings were included, but these were omitted because Yellowheads do not raise second broods and seldom attempt to renest.

Synchronization of nests within a population should be favored by all factors that reduce success of nests started later than the average but should be opposed by direct and indirect intraspecific competition among breeding birds for food, which might give advantages to later nests if there are rapidly renewing food supplies that cannot be depleted by earlier breeders. Indeed, since emergence of aquatic insects, which constitute the most important food of nestlings, continues at a high level through June and July, great variance in starting times of nests might be expected. There is no reason to expect any differences between Redwings and Yellowheads, however, as they are for the most part using the same food resources and should be subjected to the same pressures for synchronization or lack of

154

it. As shown in Table 5.6, there is little difference between the two species in variability of starting times of the first nests of the breeding females. There is, however, an important source of error in these data because we do have evidence that some female Redwings starting later nests had not attempted an earlier nest on the same or nearby marshes. If they had not nested anywhere previously that season, attributing all later nests to renesting underestimates variability in timing among Redwings.

5.4. CLUTCH SIZES

Natural selection for avian clutch size has been studied more intensively than almost any other ecologically significant trait, but molding of clutch size is very complex and is still not well understood (Cody, 1971; Fretwell,

TABLE 5.6. Timing of nests believed to represent first attempts of female blackbirds at Turnbull National Wildlife Refuge and the Potholes.

Location	Year	Mean date of first egg	Standard deviation (days)	No. of nests
		Redwing		
Turnbull	1962	16 May	9.6	45
	1964	19 May	6.6	44
	1965	15 May	6.5	21
	1968	23 May	12.6	35
Potholes	1963	22 May	6.6	11
	1964	17 May	8.5	71
	1965	5 May	9.2	27
Yearly average		17 May	8.5	
		Yellowhead		
Turnbull	1964	21 May	6.3	56
	1965	14 May	5.5	29
	1968	22 May	8.7	21
Potholes	1964	21 May	7.4	100
	1965	11 May	8.3	18
	1968	17 May	6.4	50
Yearly average		18 May	7.1	

Bowen and Hespenheide, 1974; Lack, 1954b; Payne, 1974). At this point my interest is in causes of variation in clutch size and not in determinants of means. Several possible causes of intrapopulational variation are (a) ability to mobilize energy for the clutch; (b) ability to care for offspring; (c) ability to predict foraging conditions when young are to be fed; and (d) length of breeding season and extent of renesting.

In early spring, blackbirds in Washington forage in undefended uplands, where foraging sites are probably equally available to all birds. However, there may be substantial individual variation in the rate at which prey are encountered, due to prior experience with the area, improvement of foraging skills with age, and the condition of the bird at the end of winter. One might expect older birds to be able to accumulate their energy reserves faster than younger ones, and they are also more likely to have had prior experience with the foraging area. On the other hand, individuals that bred the previous year might have been sufficiently stressed that they entered the winter in poorer condition. If so, they would require a longer time in the spring to recover to normal weight and then to accumulate reserves for breeding. Ability to care for young should also correlate positively with age.

Prediction of foraging conditions when young are to be fed requires estimates of both prey availability and densities of other users of those prey. In my study areas, the gross ranking of lakes does not normally change from year to year, barring catastrophes such as carp invasions, but there are yearly variations in overall emergence. As yet we do not know how predictable these variations are from cues available to females when they lay their clutches. However, there are reasons for believing that future food supplies should be both more predictable and more constant for Yellowheads than for Redwings. As indicated earlier, territory qualities are more constant for Yellowheads, and there are also fewer birds of other species exploiting their favored foraging areas. Redwings encounter both more variable territories and a wider variety of potential competitors. These factors should lead to a greater variation in clutch sizes in Redwings than in Yellowheads. There is no reason to expect differences between the two species in variability caused by age and skill differences.

A potential source of variability in clutch size derives from the decreasing probability of success in nesting during the breeding season. Thus, it might pay to start laying earlier with less reserves by laying a smaller clutch. This would be favored only if the probability of success of a slightly later nest were sufficiently lower to more than compensate for the reduced clutch size. Though nest success does decline during the season in Washington, it does so very slowly, especially early in the season. In addition, earlier clutches in Redwings tend to be larger than later clutches, not the reverse as would be expected if earlier breeding were favored at the expense of clutch size (Table 5.7). Greater variability in clutch size among Redwings is also expected because of the longer breeding season, the greater proportion of females that lay second clutches and the greater propensity of Redwings to replace lost clutches and broods.

TABLE 5.7. Seasonal trends in Redwing clutch sizes.

Location	Early clutches		Late clutches	
	No.	Mean clutch size	No.	Mean clutch size
TNWR, 1968[a]	18	3.8	7	2.6
TNWR, 1964[b]				
(Mann Lake)	21	4.1	11	3.6
Potholes[c]	55	4.2	18	4.2

[a] early = before June 1.
[b] early = before May 15.
[c] early = before May 18.

Clutch size variations could also be influenced if larger than average clutches were severely penalized by virtue of foraging conditions or higher predation rates on nests with more young. The extensive data of Holm (1973) for Redwings at Turnbull indicate that this type of selection was relatively weak. Number of young fledged (Table 5.8) and predation rates (Table 5.9) are the same on nests with clutches of 3, 4 and 5. However, these data do not consider the probability that a second clutch might be attempted or the probability that the female dies before the next breeding season.

Available data on Yellowhead clutch sizes are summarized in Appendix L, while Redwing data are summarized in Appendix M. As expected, variations in Redwing clutch size within any given year are greater than in Yellowheads, both on my study areas and elsewhere. Wherever Yellowheads have been studied, clutches of four strongly predominate. This is also the most common clutch size in Redwings but, in some areas, smaller clutches outnumber those with four eggs, and the range of clutch sizes is generally greater.

At the Turnbull and Columbia refuges, the influences of interlake variations in clutch size on this pattern of variability can be assessed (Tables 5.10 and 5.11). Among Redwings, yearly differences in mean clutch sizes are due to the fact that clutch sizes changed in the same direction on different lakes, while among Yellowheads clutches varied

TABLE 5.8. Relationship between clutch size and number of young fledged in Redwings. Turnbull National Wildlife Refuge, 1965, 1966, 1967 (data of C. R. Holm).

Clutch size	No. of nests	No. of young fledged/nest
3	113	1.00
4	238	0.95
5	45	0.96

TABLE 5.9. Relationship between predation rates and clutch sizes in Redwings. Turnbull National Wildlife Refuge, 1965, 1966, 1967 (data of C. R. Holm).

Clutch size	Total nests	No. of nests predated	% predated
3	113	57	50.4
4	238	123	51.7
5	45	20	44.4

more independently among lakes. This may reflect the greater importance of foraging on the marsh for female Yellowheads.

5.5. VARIABILITY IN FOODS AND FORAGING

The kinds of prey delivered to nestlings are the result of many different decisions concerning patches and prey types within patches. In addition, my multiple samples from individual nests permit me to test predictions con-

TABLE 5.10. Interlake variations in mean clutch sizes among Yellowheads. (Number of clutches in parentheses.)

Lake	Mean clutch size			
	1963	1964	1965	1968
TNWR				
Kepple		3.59 (17)		
30 Acre		4.00 (2)		
Isaacson		4.00 (17)		
Big McDowell		3.84 (10)		
Lower Turnbull		3.83 (23)	3.78 (18)	
AVERAGE		3.81 (69)	3.78 (18)	
CNWR				
Lyle		3.60 (30)		
Herman		4.00 (21)	4.04 (25)	3.61 (18)
Pit	3.63 (8)	3.33 (3)		
Hampton		3.55 (9)	3.58 (12)	3.70 (37)
AVERAGE	3.63 (8)	3.71 (63)	3.89 (37)	3.67 (55)
Cariboo Parklands				
Sorenson		3.22 (18)		
Rush		3.50 (44)		

Table 5.11. Interlake variations in mean clutch sizes among Redwings. (Number of clutches in parentheses.)

	Mean clutch size						
Lake	1962	1963	1964	1965	1966	1967	1968
TNWR							
Kepple	3.00 (4)						
Beaver	3.84 (19)		4.00 (7)		3.71 (14)	3.79 (32)	3.36 (11)
30 Acre	4.00 (3)				3.92 (22)	4.00 (23)	
Blackhorse				3.85 (13)			
Mann			3.72 (44)	4.02 (41)	3.59 (44)	3.87 (40)	3.63 (16)
Little McDowell	3.00 (2)			4.06 (51)	3.50 (48)	3.53 (30)	3.86 (7)
Isaacson	3.22 (9)			4.00 (4)			
Lower Turnbull	4.25 (4)			3.72 (11)	4.28 (7)		
Other	3.50 (12)						
AVERAGE	3.62 (53)		3.76 (51)	3.98 (120)	3.63 (135)	3.84 (125)	3.59 (34)
CNWR							
Lyle		4.29 (7)	4.00 (15)	4.41 (17)			3.65 (26)
Herman			4.20 (5)				
Pit		4.17 (6)					
Willow			4.23 (13)	4.18 (17)			3.92 (13)
Shoveller			4.12 (25)				
Other							3.31 (13)
AVERAGE		4.23 (13)	4.12 (58)	4.29 (34)			3.64 (52)

cerning the extent of *individual* differences in foraging behavior among Redwings and Yellowheads. The most difficult problem is to decide which interspecific comparisons are most meaningful and why differences are or are not to be expected. In particular, I wish to determine whether or not there are greater individual differences among Redwings exploiting a given marsh than among Yellowheads in the same or comparable marshes.

My expectations are based on the fact that all prey regularly taken by blackbirds appear to be readily available to both species. There are no prey taken by the largest (male Yellowheads) that are not easily handled by the smallest (female Redwings) birds, and there is a high within-habitat overlap in foods taken by foraging adults (Chapter Six below). Evidently, morphological and behavioral differences between species are not great enough to cause significant differences in prey encountered or in the ease with which different prey can be captured. In fact, their bill measurements are more similar than their weights and tarsal measurements, suggesting that morphological differences relate primarily to where foraging takes place rather than what can be captured.

Nonetheless, though the amount of morphological variation *within* species is certainly insufficient to favor differential prey selection behavior or to generate significant differences in frequencies of encounter with prey types, it does not follow that morphological differences are not sufficient to affect foraging efficiencies in different patch types, thereby leading to differential use of available patches.

For several reasons Redwings forage in a greater variety of patches than Yellowheads. First, on marshes where both species breed, Redwings are displaced to the peripheries where foraging opportunities close to nests are poorer than on Yellowhead territories. Lakes with only Redwings are lower-productivity lakes, which has the same effect. In both cases, the unavailability of high-quality patches favors the use of more patch types. Also, differences in effective patch quality should be less for Redwings, because high quality, undefended edge is usually found farther from their nests. Yellowheads, on the other hand, typically have high-quality foraging close to their nests, and low-quality patches are distant.

Second, Redwing nests are closer to a wider variety of patches than are Yellowhead nests. This increases the probability of use of low-quality patches, as little travel is required to reach them (see Chapter Four above). Moreover, many of these patches are terrestrial and, hence, more varied in structure than the emergent vegetation that normally extends some distance around Yellowhead nests.

Third, the smaller size of Redwings appears to increase their efficiency of foraging in trees which are regularly used whenever they are available. In fact, in some parts of the range, most prey fed to nestlings are taken from trees (Nero, 1956a; Snelling, 1968). Yellowheads breed primarily in regions with few trees and avoid them when they are close to their territories.

Therefore, foraging Redwings on the average encounter a wider range of patch types close to their nests, and they should use more of these patches because the best are not of highest quality. Yellowheads encounter only a few patch types near their nests, and these are usually of high quality. Accordingly, within a particular marsh, we expect greater differences among female Redwings in the food delivered to their young, reflecting primarily differential availability of patch types close to the nests.

To test predictions concerning the extent of individual differences in foraging I assembled food data from all nests for which I had at least four food samples (four hours of delivery). In most cases, this amounts to more than 100 prey items per nest. For one analysis correlation coefficients were determined between the numbers of prey in each food category between two nests for samples from simultaneously active nests on the same lake. Each nest was compared to all other nests and a mean value for the amount of similarity among nests was calculated for the marsh (Table 5.12).

As data on deliveries to individual nests are available for only Yellowheads at the Potholes, interspecific comparisons can be made only at Turnbull. The most interesting comparison is between Yellowheads on Kepple Lake and Redwings on Beaver Pond, because most Redwings with nestlings on Beaver flew to Kepple where they foraged on undefended edge not far from the emergent vegetation supporting Yellowhead nests. Individual female Redwings were more different from one another than were individual female Yellowheads (0.57 vs. 0.72), but I cannot distinguish more variable selection of prey on the edge habitat on Kepple from differential use of aquatic habitats on Beaver. The latter sites would have been, on the average, of much poorer quality (as emergences are so low on Beaver) and would have offered a different array of prey types than

TABLE 5.12. Correlation coefficients between number of prey in each food category between two nests for samples from nests on the same marsh. Data only from nests from which at least four samples were obtained.

Year	Lake	No. of nests	Correlation coefficient $\bar{X}$ (r)
	Yellowhead		
1964	Herman, CNWR	5	.57
1964	Kepple, TNWR	5	.72
1965	Lower Turnbull, TNWR	5	.84
	Redwing		
1964	Mann, TNWR	5	.78
1965	Beaver, TNWR	5	.57

those on Kepple. Results from Redwings on Mann Lake and Yellowheads on Lower Turnbull can also be tentatively compared, because Redwings fed primarily on McDowell Lake, which is similar in productivity to Lower Turnbull. In this case, individual differences in prey delivered to the nests are similar between the two species.

An alternative way of analyzing the same data on deliveries to specific nests is to compare diversity of prey brought to each nest with overall diversity of prey delivered to nests on that lake. If overall breadth of food delivered is the result of foraging by a number of relatively distinct but specialized individuals, then mean diversity of prey delivered to each nest should be much lower than the prey diversity of the pooled samples. If, on the other hand, overall breadth is caused by foraging by similar but generalized individuals, each exploiting the full range of prey types, then mean diversity of prey delivered to each nest should be similar to prey diversity of the pooled samples. The results of this analysis are given in Table 5.13. Statistical comparison of the data is impossible, as there are only three diversity values for Yellowheads and two for Redwings, but individual female Redwings may be slightly narrower in their foraging relative to the population as a whole than female Yellowheads. For both species, however,

TABLE 5.13. Diversity of prey delivered to individual blackbird nests compared to prey diversity of the combined nest samples. ($H = -\Sigma p_i \log_{10} p_i$)

Year	Lake	Mean diversity of prey delivered to individual nest	Overall prey diversity	Overall div./ mean div.
		Yellowhead		
1964	Kepple, TNWR	.431	.475	.907
1964	Herman, CNWR	.794	.975	.814
1965	Lower Turnbull, TNWR	.805	.908	.887
		Redwing		
1964	Mann, TNWR	.800	.998	.802
1965	Beaver, TNWR	.848	1.014	.836

the differences among individuals are not striking, individuals averaging more than 80 percent as broad as the population as a whole. Therefore, I tentatively reject the hypothesis that there are major differences between the two species in relative extent of within- and between-phenotype contributions to overall breadth of foods and foraging.

The only other information I have concerning individual differences in foraging were obtained in 1965 on Lower Turnbull Lake. Though I was not sampling emergence quantitatively that year, it was obvious that the normal heavy emergence of odonates was not occurring. Instead of flushing clouds of teneral damselflies during the emergence period, I disturbed no more than a few individuals per step. Correlated with this was an unusually high nestling starvation rate. No nest fledged more than three young, and only five nests out of twelve that I was watching fledged three. In other nests the original three or four nestlings were reduced to two by the time they were one week old. Comparisons of prey delivered to nests fledging three vs. two young are given in Table 5.14. The data suggest that female Yellowheads who fledged only two young continued to search for damselflies despite their lower availability. These individuals brought odonates to

TABLE 5.14. Food delivered to nestling Yellow-headed Blackbirds at nests fledging 2 vs. 3 young during a year of failure of the damselfly emergence. Lower Turnbull Lake, Turnbull National Wildlife Refuge, June 1965.

	Nests fledging only 2 young	Nests fledging 3 young
Total hours of sampling	37.6	15.1
Insects delivered/nest/hr	22.8	24.3
Percentage of odonates	19.2	16.5
Percentage of samples with odonates	80.6	60.0
Odonates/nest/hr	4.37	4.04
Percentage of dytiscids	9.4	20.3
Percentage of samples with dytiscids	69.4	86.7
Dytiscids/nest/hr	2.1	4.9

the nests most of the time (80.6 percent of samples) but did not achieve high rates of delivery (4.4 odonates per hour per nest). Female Yellowheads that fledged three young either intrinsically had different foraging modes or changed their foraging pattern to search more intensively for other kinds of aquatic prey, especially dytiscid larvae, which are taken from the water rather than from emergent vegetation as are damselfly tenerals. These individuals brought odonates to their nests less often (60 percent of samples) but brought dytiscid larvae more often (86.7 percent of samples compared to 69.4 percent of samples). A higher fraction of the damselflies delivered to nestlings in nests fledging only two young were tenerals (31.3 percent vs. 10.2 percent) while adults constituted a higher proportion of damselflies delivered to nestlings in nests fledging three young (83.0 percent vs. 58.0 percent). However, the proportion of naiads was low for both groups (10.7 percent in nests fledging two young; 6.8 percent in nests fledging three young) which does not indicate that they were searching in the water.

Evidence for species differences in overall breadth of foods taken is more extensive than for individual differences. Analysis is made somewhat simpler by the fact that breeding seasons of blackbirds in the Pacific Northwest are relatively short, and most of my food sample data were gathered within a period of six weeks. Within that interval there is little evidence of seasonal shifts in prey delivered. My most extensive data on season changes were obtained with Yellowheads on Lyle Lake in the Potholes in 1963, where large samples were obtained regularly over a four-week period. During this time there was little change in prey delivered to nestlings (Figure 5.2). This is not surprising, because the peak of emergence of aquatic insects occurs later in the season than the peak in blackbird breeding, and termination of breeding is clearly not set by

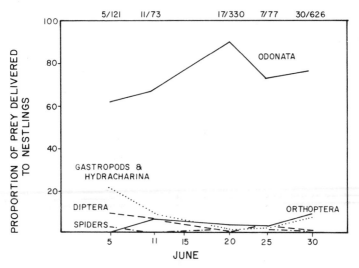

FIGURE 5.2. Seasonal pattern of prey delivered to nestling Yellowheads, Lyle Lake, Potholes, 1963.

availability of high-quality food for nestlings (see Chapter Two). Therefore, in subsequent analyses, I ignore seasonal changes in foods and foraging.

Because of problems of assigning prey to patch types and because my estimates of use of different patch types are necessarily based primarily on inference from prey delivered to nestlings, I can compare patch use only at the Potholes. Criteria for assigning prey to patch types are detailed in Orians and Horn (1969). Basically, from collections of prey in different patch types we determined which prey were almost always found within a single patch type. Prey not unambiguously assignable were placed in patches on the basis of prey with which they were associated within a particular sample. Thus, a prey type regularly found in both sage and along edges of ponds was assigned to the edge if it was associated with prey normally restricted to edges but assigned to the sage if other prey in the sample indicated foraging in sage.

167

At the Potholes the three species of blackbirds differ in the extent to which they use different patch types. The most striking difference is the virtual exclusion of Brewer's from emergent vegetation with the result that the proportion of its prey taken at sedge-grass edges of ponds is higher than among Redwings and Yellowheads (Orians and Horn, 1969). Nonetheless, the overall diversity of use of patch types does not appear to be significantly different among the species (Table 4.7). There are some interesting differences in the diversity of prey types taken within different patches, which may reflect the ease with which individuals of the three species are able to forage in those patch types. For example, while Redwings and Yellowheads do not differ in the diversity of prey they take while foraging in emergent vegetation, they both seem to take less diverse prey at sedge edges than do Brewer's, which are best adapted morphologically for foraging in that patch type (Table 5.15). Similarly, Yellowheads, which are less adept at foraging among branches of shrubs and trees, took less diverse prey when foraging in sagebrush than did Redwings and Brewer's. Without direct observations of the details of foraging and prey capture these differences are only suggestive, but we do know that when foraging in

TABLE 5.15. Diversity of foods taken in different patches by blackbirds gathering food for nestlings in the Potholes, 1964-1965 [$1/(\Sigma p_i^2)$]. Complete data are given in Orians and Horn, 1969.

		Time of day			Time-weighted average
	Species	0400-0700	0700-1300	1300-1900	
Emergent vegetation	YH	2.6	2.4	2.8	2.6
	RW	2.9	1.8	2.8	2.4
Sedge edge	YH	2.4	1.3	1.1	1.4
	RW	-	2.0	1.5	1.8
	BB	1.4	2.4	2.8	2.4
Sagebrush	YH	2.2	1.4	2.9	2.2
	RW	3.2	3.4	3.3	3.4
	BB	4.5	2.3	3.1	3.1

sagebrush Yellowheads captured primarily aquatic prey and took few lepidopteran larvae, whose capture requires careful search within the branches of the shrubs.

Though prey could not be assigned with certainty to patch types at Turnbull, some useful comparisons can nonetheless be made (Table 5.16). Except for Kepple Lake, Yellowhead food samples contain more prey types than Redwing food samples, but in all cases sample sizes for Yellowheads are much larger than for Redwings. Because rare prey are added with increasing sample sizes, this comparison is not particularly meaningful. Standard measures of diversity are insensitive to rare prey, which makes comparisons of diversity of prey taken on different lakes more meaningful. For the five lakes at Turnbull for which samples from both Redwing and Yellowhead nests are available, diversity of prey delivered to Redwing nests was greater at four of them, and the overall mean diversity of prey delivered to nestling Redwings (0.816) was much greater than that delivered to nestling Yellowheads (0.549). Also, in five of six cases, the percentage of prey that were from aquatic taxa was higher for Yellowheads than Redwings. In the sixth lake, there was no difference.

Another comparison of interest is the similarity in prey delivered to nestlings by birds at the same sites but in different years (Table 5.17). Yellowheads, being confined to the most productive lakes, should be especially sensitive to changes in productivity. Because they settle at high densities on those lakes, they might be forced to forage at great distances from their nests and to exploit suboptimal patches extensively during poor years. Yellowheads breeding on Herman Pond in 1968, a year with a heavy carp infestation and almost no emergence on Herman itself, did most of their foraging at Coot Lake, about 1.1 kilometers from their nests. Redwings, because they are pushed to the poorer sites anyway, exploit a range of habitats every year

TABLE 5.16. Comparisons of number of prey types, prey type diversity ($1/\Sigma p_i^2$) and percentage aquatic prey taken by Redwings and Yellowheads on different lakes at the Turnbull National Wildlife Refuge. (*n.c.* = not calculated because sample size is too small.)

Lake	Comparison time	No. of prey		% aquatic		Diversity		No. of prey types	
		RW	YH	RW	YH	RW	YH	RW	YH
Blackhorse	total	156	301	.846	.937	.587	.602	13	15
	preemerg.		161		.901		.386		8
	emergence		78		.987		.453		6
	postemerg.		62		.968		.445		7
Kepple	total	118	1359	.593	.995	1.040	.124	18	12
	emergence		1086		.999		.081		8
	postemerg.		273		.978		.249		9
30 Acre	total	129	683	.558	.832	.889	.755	12	16
	preemerg.		129		.977		.684		8
	emergence		330		.715		.785		15
	postemerg.		224		.920		.782		13
McDowell	total	118	405	.839	.832	.620	.556	12	17
Isaacson	total	31	552	.613	.801	*n.c.*	.683	9	17
	preemerg.		98		.592		.799		14
	emergence		264		.837		.654		13
	postemerg.		160		.869		.483		15
Lower Turnbull	total	89	3894	.562	.909	.944	.707	14	20
	preemerg.		254		.882		.570		12
	emergence		1773		.947		.630		19
	postemerg.		1867		.877		.721		18
Mann	total	663		.195		.957		17	
	preemerg.	217		.272		.993		12	
	emergence	214		.215		.945		15	
	postemerg.	232		.103		.807		13	
Beaver	total	1066		.741		.829		19	
	preemerg.	229		.930		.478		13	
	emergence	535		.914		.404		14	
	postemerg.	302		.265		1.109		19	

TABLE 5.17. Yearly differences in diversity of foods delivered to nestling blackbirds, calculated using Pianka's (1973) index of overlap (α_{ij}).

Comparison	Overlap
Yellowhead	
Rush Lake 1963 vs. Rush Lake 1964	.792
Rush Lake 1963 vs. Westwick Lake 1964	.062
Rush Lake 1964 vs. Westwick Lake 1964	.629
CNWR 1963 vs. CNWR 1964	.975
CNWR 1963 vs. CNWR 1965	.345
CNWR 1964 vs. CNWR 1965	.497
Redwing	
Mann 1964 vs. Beaver 1965	.420
CNWR 1964 vs. CNWR 1965	.963
Taboga 1966 vs. Taboga 1967	.801
Yellowhead - Redwing	
YH, CNWR 1964 vs. RW, CNWR 1964	.662
YH, CNWR 1965 vs. RW, CNWR 1965	.512

and, in the event of emergence failure on the best lakes, their foraging patterns should be less affected.

For example, in 1963, a good emergence year in the Cariboo Parklands, there was almost no overlap ($\alpha_{ij} = 0.062$) between prey delivered to nestling Yellowheads at Rush Lake (a productive lake) and Westwick Lake (a poor lake). In 1964, a poorer year, there was much more overlap ($\alpha_{ij} = 0.629$), but that was still less than the overlap at Rush Lake between the two years ($\alpha_{ij} = 0.782$). A direct comparison between Redwings and Yellowheads is available for the Potholes in 1964 and 1965. Carp infestations were already having their effects in 1965 and, as predicted by the argument just advanced, there was less overlap between prey taken by Yellowheads in the two years ($\alpha_{ij} = 0.497$) than in prey taken by Redwings ($\alpha_{ij} = 0.963$). Other comparisons shown in the table are of dubious value.

5.6. CONCLUSIONS

In general, the results of this chapter indicate that Redwings and Yellowheads respond similarly to changes in availability of various environmental resources and that differences between their patterns reflect primarily the differential availability of resources in the vicinity of their nests. There is no evidence that there is any more genetic variability related to phenotypic traits analyzed here in one of the species compared to the other, but my tests are all indirect and the question remains an open one. Redwings are more variable than Yellowheads where they settle, the nest sites they use and in their clutch sizes. If I am correct that birds of both species rank patches similarly, then differences in availability are due mostly to (a) direct competitive interactions among the birds, especially interspecific territoriality, which exerts a strong effect on the patches available for use by Redwings, and (b) the smaller size of Redwings, which influences the range of habitats they can use. Such differences and the interactions related to them affect the structure of the community in ways that will be explored in the next chapter.

CHAPTER SIX

The Patterns: Competition, Overlap and Community Structure

Though there has been little agreement among ecologists over the importance of competition in molding characteristics of ecological communities, owing mostly to the scarcity of direct evidence for or against competition (Orians and Collier, 1963), workers with many different groups have been convinced that competition occurs regularly enough to be a significant proximate and ultimate factor influencing community structure. Birds are one such group (see Cody, 1974, for an extensive recent review of the evidence and its interpretation). For the most part, however, competition theory was developed and predictions were made from it without being based on any underlying theory of resource exploitation by individuals comprising competing populations until MacArthur (1969, 1972) formally incorporated these notions into the equations. Recent developments in optimal foraging theory provide additional bases for predicting proximate responses to reduced prey availability, whether or not it is due to competition. Extensions can also be made to longer-term consequences.

In the following treatment I will be concerned primarily with direct and indirect competition among blackbirds for food during the breeding season and its influence on foraging behavior and prey and habitat choices. I have insufficient data to assess competition at other times of the year or under other circumstances. I use concepts and terminology from optimal foraging theory as developed in Chapter Four as the basis for predictions about what kinds of changes in patch and prey choices might be expected if blackbirds compete. Evidence that they may compete comes from the existence of nestling starvation, interspecific territoriality and extensive overlap in patch use and prey taken. I consider, in turn, the long-term effects of competition, competition in different types of foraging patches, and the significance of size differences on the outcome of competitive interactions.

6.1. SHORT-TERM EFFECTS OF COMPETITION

The proximate effects of competition may vary considerably with the kinds of competitors and the types of environments in which they interact. It is useful to consider four general cases of competitors in a single patch environment.

CASE I. TOTAL OVERLAP: IDENTICAL RANKING OF PREY BY PREDATORS

This might be true for obligate specialists or for generalists whose morphology and behavior are so similar that their efficiencies of capture of all available prey are the same. Foraging by two such competitors in the same patch should result in a lowering of encounters with *all* prey utilized by both and, hence, a lowering of net energy intake. Both predators should then expand their diets, but, as rankings are identical, new prey are added in the same order by both of them. For a given reduction in net energy intake, generalists should expand their diets more than specialists, because the next-ranked prey, not previously in the optimal set, is not as different from previous prey as for a specialist. In both cases, however, the overlap remains complete whatever the intensity of prey depression.

174

CASE II. PARTIAL OVERLAP: SIMILAR
RANKING OF PREY BY PREDATORS

Ecologically similar closely related species would normally fall into this category. Sympatric foraging by two such competitors lowers availability of some prey more than others, but high-ranking prey are likely to be particularly depressed because they are preferred by both predators. Both predators should expand their diets in response to lowered prey availabilities, and new prey types may be largely, though not completely, shared. The result may be little change in the amount of overlap among the two predators as prey are depressed. However, if the predators have different adaptations for capturing less desirable prey, overlap may decrease with decreasing abundance of preferred items.

CASE III. PARTIAL OVERLAP: DISSIMILAR
RANKINGS OF PREY BY PREDATORS

This case should be common among predators that forage in similar ways but are very different in size so that the larger predator can use prey unavailable to the smaller. Large prey are likely to have high ranks for all predators but, as prey approach upper limits of the size that can be attacked, prey rank should drop abruptly because of greatly increased handling times (Schoener, 1971) or because the prey is dangerous to attack. This case might also apply to predators that take similar prey but employ different foraging behaviors and morphological equipment.

This category contrasts with the previous one because a competitor depresses preferred prey relatively little compared to lower-ranking prey. Nevertheless, net energy intake per unit time should drop for both predators, though less than for Case II, and both should be expected to expand their diets. In the case of predators differing in size the newly added prey are likely to be small and may often constitute a small energetic fraction of the diet. As a result, the percentage of prey types taken by both predators may increase while actual dietary overlap (energy content) decreases.

CASE IV. SPECIALIST COMPLETELY OVERLAPPING WITH GENERALIST

This varied category potentially includes many real cases of competition because most interactions involve species with different K's and different niche breadths. Also, the immediate effects of arrival of a competitor depend on the position of the specialist with respect to resource utilization by the generalist. A specialist may, for example, prefer the top-ranked or lower-ranked prey of a generalist. Whichever prey it prefers can be expected to be depressed, but depression of a low-ranked prey is less serious for the generalist than depression of a high-ranked prey, unless the former is very abundant and the latter very rare. Net energy intake per unit time should be lowered for both predators and both should expand their diets, the generalist more so than the specialist unless the specialist prefers a prey that is very low ranked and uncommon for the generalist.

Marsh-nesting blackbirds appear to fit Case II most closely. They are similar in morphology and hunting behavior and, for prey utilized during the breeding season at least, they handle available prey with similar ease. The largest prey (dragonflies, cicadas) I have found in food samples are taken by all three species, and there are no apparent interspecific differences in handling times for large prey. If assimilation efficiencies are similar, prey items should have similar ranks for all three species.

Assuming that blackbirds are examples of Case II, then

the proximate effect of interspecific competition should be a general reduction of prey, especially of high-ranked types. Therefore diets should expand for all three species when competition for food increases, provided there is no change in prey species due to other causes. As indicated in Chapter Four, there is evidence that food is absolutely most abundant and concentrated in mid to late morning in marshes in western North America. Because blackbirds forage actively all day when they have nestlings, overlap might be greater at that time of day than earlier or later, and dietary contraction is to be expected then. However, as encounter rates with large prey are seldom high enough to cause rejection of smaller prey at this time (see Chapter Four), direct predictions about changes in prey overlaps cannot be made. Therefore we cannot unambiguously infer competition from extent of overlap. Nevertheless, patterns in dietary overlap do reveal something about foraging similarities and potential competition.

Though the three blackbirds I have studied most intensively are similar in size and manner of foraging, there are important differences that *do* influence their foraging efficiencies in different microhabitats. Compared to Redwings and Yellowheads, Brewer's are more slender and have longer legs (Table 6.1). In addition, they have smaller muscles for opening their slender bills against resistances (Beecher, 1942). These morphological differences have important consequences for foraging modes and efficiencies. For example, Brewer's rarely dig in turf for hidden prey and infrequently use their bills in gaping movements to flip over sticks, rocks, cow pats and so on in their search for food. These techniques are utilized heavily by both Redwings and Yellowheads. Also, in sagebrush bushes Brewer's forage primarily by leaping up from the ground or walking on top of the shrubs while Redwings regularly climb through them.

Table 6.1. Morphological measurements (mm) of Redwings, Yellowheads and Brewer's Blackbirds (data from Ridgway, 1902).

Location	Sex	No. of specimens	Wing	Tail	Culmen	Bill depth	Tarsus
Yellowhead							
Utah, California	♂	6	142.0	103.9	22.9	12.2	24.6
	♀	5	113.3	82.3	20.6	10.4	30.7
Redwing							
Great Basin	♂	11	123.4	91.2	23.9	11.9	30.5
	♀	7	99.6	72.1	19.6	10.7	26.9
Northwest Coast	♂	9	123.2	91.7	24.4	11.7	29.5
	♀	9	103.1	77.5	20.8		19.1
Brewer's							
Rocky Mountains	♂	5	131.1	101.1	23.1	9.9	32.5
	♀	4	118.9	89.1	19.8	9.1	31.0

Redwings and Yellowheads also scratch frequently when foraging for seeds. The substrate is moved by a vigorous backward movement of both legs, the left invariably preceding the right, though a stronger kick is delivered by the right foot. Interspersed with scratching, they also dig holes, sometimes quite deep, with their bills. When foraging in this manner, Redwings and Yellowheads move slowly over the ground and often spend time in one spot digging and scratching to reveal otherwise hidden food. Brewer's, on the other hand, move at a much more rapid and steady rate, seldom digging in the turf or scratching.

The longer legs of the Brewer's permit them to forage with greater ease among grasses and sedges on lake edges at the Potholes, but they are very awkward when attempting to forage on vertical stalks of cattails and bulrushes. Apparently they are unable to turn their legs to straddle between two adjacent stalks as do foraging Redwings and Yellowheads. When Brewer's forage in marshes, as they do regularly at Turnbull, they prefer blown-down cattails and bulrushes and walk over the more or less flat surfaces.

In an attempt to measure comparative blackbird foraging efficiencies among cattails I constructed artificial marshes by impaling cattails on large nails driven through a 2m × 2m piece of 1.9 cm plywood. The plywood was weighted down in approximately 25 cm of water in an aluminum basin 2m × 2m × 0.3m. Two densities of cattail stalks were employed in the experiments, high density (440 stalks = 110/m²) and low density (100 stalks = 25/m²). Using a table of random numbers, 80 stalks in each marsh type were selected as sites for prey replacement. The prey, mealworms, were attached to the stalks by impaling them on headless insect pins. In the dense marsh there were so many cattail stalks that placement of prey in the marsh center was very difficult. Consequently, 96 center stalks were not used, and the 80 mealworms were distributed over the remaining 344. Blackbirds readily spotted the mealworms and removed them from the pins with no difficulty. No mealworms were placed close enough to the edge to permit birds to capture them from the basin rim but, especially in low cattail density experiments, it was possible for the birds to see into most of the marsh by walking around the edge. This affected the rate at which prey were captured, because the birds, once they had learned the experimental design (which they did very rapidly), usually foraged by searching from the rim and entered the marsh only to capture a previously located prey item.

In each experiment four birds, all of which had been in captivity at least four months and had lived in the cage with the marsh, were permitted to forage for five minutes once the mealworms were in place. Timing began when the first bird took a mealworm, which usually occurred almost instantly when the experimenter left the aviary. In fact, toward the end of the series of experiments with Brewer's, it was necessary to employ two experimenters, one to install the mealworms and the other to prevent the birds from beginning to forage on the opposite side of the marsh before all prey had been placed! This problem was particularly acute with Brewer's, probably because all birds were maintained on a seed diet between experiments, and Brewer's seemed much more highly motivated to capture insects than the more granivorous Redwings and Yellowheads. In fact, I believe that the stronger motivation of the Brewer's caused results the opposite of those predicted (Table 6.2).

In both types of marsh, Brewer's took a larger number (and proportion) of mealworms from both center and edge portions of the marsh during a standard five-minute experimental period, even though it was obvious that they had difficulty in moving through the cattails and often got wet because they fluttered into the water when going for the prey. Redwings and Yellowheads moved easily through the cattails, but in most experiments none of them were foraging in or around the marsh after the first couple of minutes. The small percentage of prey taken does not reflect the ability of the birds to capture them but rather their interest in doing so. That the Redwings and Yellowheads were capable of much higher rates of foraging was demonstrated by three experiments in which two males of each species were used. In these experiments most prey were taken within a few minutes, the birds apparently being stimulated to forage by the presence of individuals of the other species (Table 6.2).

A final behavioral difference important in determining foraging locations of blackbirds is interspecific dominance relations. Redwings and Yellowheads are interspecifically territorial, and when tested at grain bait stations on neutral ground far from any territories at Turnbull, Yellowheads were always dominant to Redwings (Orians and Willson, 1964). Brewer's Blackbirds are subordinate to both Redwings and Yellowheads on neutral ground and are not permitted to forage on territories of either species. Therefore, Brewer's are restricted in their foraging to neutral

TABLE 6.2. Number of mealworms captured by blackbirds in an artificial marsh during a 5-minute experiment involving four birds (see text for details). (P = proportion)

| | Dense marsh | | | | | Loose marsh | | | | |
| | Center | | Edge | | | Center | | Edge | | |
Species	No. of exp.	No.	P	No.	P	No. of exp.	No.	P	No.	P
Brewer's ($\delta + \varphi$)	23	10.9	.137	27.4	.343	12	8.5	.108	28.7	.358
Redwings (φ)	10	6.8	.085	20.4	.257	13	4.9	.061	18.5	.232
Yellowheads (δ)	-					15	2.6	.035	24.5	.307
Redwings (δ) + Yellowheads (δ)	-					3	35.3	.803	32.7	.907

181

areas. Most of the uplands on my study areas are undefended, but on edges of lakes and streams, only areas without emergent vegetation that supports territories of Redwings and Yellowheads are accessible to Brewer's.

Differences in dominance and morphology among the
blackbirds probably determine, in the proximate sense,
breeding distributions of the three species and locations of
nests and foraging sites. The theory of Central Place Foraging can be adapted to deal with this problem. Consider, for
simplicity, the case in which a predator forages until it captures a fixed quantity of food (L_i) in patch i and then returns to the nest.

Let: $g(D_i, L_i)$ = time to get L_i food at prey density D_i and
T_{T_i} = time to travel to and from patch to nest. Then the
rate of delivery of food from patch i to the nest is

$$\frac{L_i}{T_{T_i} + g(D_i, L_i)}.$$

As L_i is fixed, the problem to the bird, if maximizing the
rate of delivery of energy to the nest is appropriate, is one
of minimizing $T_{T_i} + g(D_i, L_i)$. T_{T_i} is also fixed by the position of the patch relative to the nest, but $g(D_i, L_i)$ may vary.
An example involving four imaginary patches at varying
distances from a nest is shown in Figure 6.1. The dashed
lines $(T_T + T_L = C)$ connect patches yielding equal rates of
energy delivery to the nest. For example, when prey capture rates in patch B, initially the best of the four patches
shown in Figure 6.1, are reduced to the position indicated
by the * the birds should then, and only then, begin to use
patch D as well.

I apply this general model to the three blackbirds at the
Potholes in Figures 6.2 a, b and c. In all three diagrams
A = edge of cattails, B = upland dry grass and sage and
C = sedge edges. When represented with a prime, the
postemergence period is signaled; without the prime the

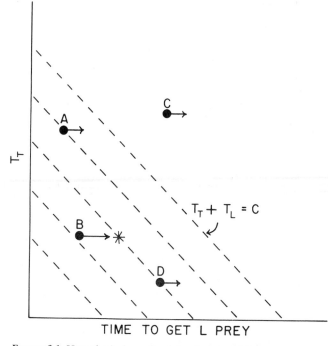

FIGURE 6.1. Hypothetical case involving four imaginary patch types at varying round trip traveling times (T_T) from a central place (nest). Patches differ in the rates at which prey are encountered $(T_L$ = time to gather a full load) but may shift in quality, i.e., move horizontally on the graph. Dashed lines connect points yielding equal rates of energy delivery to the nest $(T_T + T_L = C)$ (after Charnov, 1973).

emergence period is signaled. For both aquatic patch types, prey encounter rates are assumed to be higher during the emergence periods, while in the upland habitats the reverse is assumed. Because Brewer's nest in sagebrush bushes away from water and usually not near patches of cattails with territories of Redwings and Yellowheads, travel time to cattails is assumed to be higher than travel time to upland habitats (closest) and sedge edges (intermediate distances). Since Redwings and Yellowheads nest

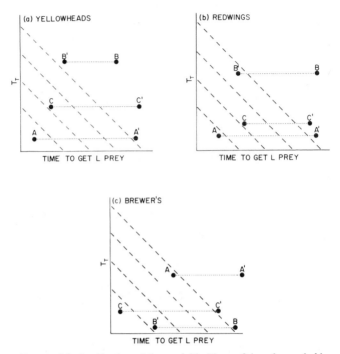

FIGURE 6.2. Application of the model in Figure 6.1 to the probable case for the three species of blackbirds at the Potholes. A = edge of cattails, B = upland dry grass and sagebrush, and C = sedge edges. (a) probable patch situation for Yellowheads; (b) probable patch situation for Redwings; (c) probable patch situation for Brewer's Blackbirds. See text for further explanation of shifts in patch quality between emergence period (without a prime) and post-emergence period (with a prime).

in the same situations, travel times are treated as equivalent.

Because of the morphological differences indicated above, Brewer's are represented as having higher rates of prey capture in sedge edges while Redwings and Yellowheads have highest rates of capture in emergent vegetation. These relationships predict preferential use of aquatic habitats by all species during the emergence period, with Brewer's concentrating on sedge edges while

Redwings and Yellowheads use both sedges and cattails. The uplands become profitable for all three species only in the postemergence period (or during the preemergence period, which is not shown).

If generally true, these relationships explain the distribution of foraging individuals of the three species. The subordinate Brewer's are most common in areas where there are rich, undefended edges of lakes and streams, which occur where edges are not occupied by beds of emergent vegetation. At the Potholes there are large stretches of such areas, and Brewer's are common. At Turnbull, however, most lakes are completely ringed by beds of cattails and bulrushes, and there are, as a consequence, almost no undefended rich foraging areas. At Turnbull, Brewer's are very scarce, being limited to the mowed and well-watered lawns around the refuge headquarters (artificial sedge edges?) and a few other sites where there are sedge meadows without cattails or bulrushes. A similar pattern occurs along the gradient from the Great Plains to the Midwest, but interpretation of this gradient is complicated by the presence of the Common Grackle (*Quiscalus quiscula*), which is also primarily a forager along sedge edges.

When foraging in the same patch type, blackbirds of all three species take similar prey (Table 6.3), except Brewer's and Yellowheads during the emergence period. Given the structural simplicity of the patches, this is to be expected. Moreover, while the morphological differences among the three species affect foraging efficiencies in different patch types, they probably do not influence how many foraging modes are effective within a particular patch type. For this reason, interspecific differences while foraging involve differential use of patch types more than taking different prey within patches, that is, food overlap is greater than overlap in patch use (Table 6.3). However, as pointed out

185

TABLE 6.3. Overlap in food and foraging patch of blackbirds gathering food for nestlings in the Potholes (from Orians and Horn, 1969).

Type of overlap	Species compared	Time of day 0400-0700	0700-1300	1300-1900	Time-weighted average
Patch use	YH-RW	.88	.97	.56	.79
	RW-BB	.93	.18	.99	.65
	BB-YH	.80	.21	.60	.48
Food (overall)	YH-RW	.73	.97	.64	.79
	RW-BB	.68	.75	.96	.82
	BB-YH	.71	.84	.66	.74
Food (within-patch)	YH-RW	.79	.93	.83	.86
	RW-BB	.72	.36	.96	.67
	BB-YH	.57	.10	.76	.46
Total	YH-RW	.72	.91	.48	.70
	RW-BB	.67	.06	.95	.54
	BB-YH	.47	.02	.46	.29

in the previous chapter, Brewer's take a wider variety of food in sedge edges, their favored patch type, while Yellowheads and Redwings take the same variety of prey in emergent vegetation and sedge edges, where they both use the same foraging modes.

6.2. LONG-TERM EFFECTS OF COMPETITION

Though the expected proximate effect of resource depression by a competitor is an expansion of diet, these changes in diet, combined with the altered abundances of prey, may change natural selection sufficiently for the average morphologies of one or more of the competitors to become less suitable than different ones. For example, if formerly high-ranking and abundant prey are scarce and contribute less to the diet, while newly added prey are common and constitute a high proportion of the diet, a different phenotype may well be optimal than under previous prey availabilities. Phenotypic changes that increase efficiency of encounter, pursuit, capture and/or handling of a new prey type may result in a net increase in energy intake per unit time spent foraging. If so, phenotypes more different from the competitor than the previous optimum phenotype should be favored, resulting in divergent evolution.

For reasons given in the previous section, phenotypic changes that adapt blackbirds more closely to the characteristics of a given patch are more likely to evolve than characteristics resulting in two or more ways of foraging efficiently within a single patch type. Foraging over water from vertical stalks of emergent vegetation requires very different movements and positions than foraging on the ground at marsh edges or in upland dry grasses. Patch types also differ in productivity, visibility and dimensionality (linear edges vs. two-dimensional patches), which affect costs and benefits of defending them. Therefore, I discuss each major patch type used by blackbirds on my study areas in terms of the optimal phenotypes for foraging there and the form that competition has apparently taken in each of them.

Competition in Emergent Vegetation and Short Grass

Patches of emergent vegetation, such as cattails and bulrushes, are readily defended because visibility is excellent, productivity is often high and the patches are usually of sufficient breadth that the optimal polyhedral shapes of territories can often be realized (Grant, 1968). Not surprisingly, North American blackbirds that nest in marshes and gather most of the food for their nestlings from them are strongly territorial. For birds the size of blackbirds, however, foraging from vertical perches is difficult for individuals with relatively long tarsi. In patches of emergent vegetation horizontal distances between perches are often short, and nearly all suitable perches are vertically oriented and many are unstable. These features put a premium on short tibia and tarsi and an ability to position legs so that they extend laterally from the body, whereas foraging on solid substrates and in an upright position favor relatively longer tarsi (Grant, 1965, 1966). Redwings and Yellowheads commonly assume such perching positions when foraging in emergent vegetation, while Brewer's apparently cannot do so.

Teneral aquatic insects are found superficially on vertical stalks, and their defensive behavior, while they are still not hard enough to fly, is to sidle around a stalk, keeping the stalk between them and the predator. Efficient capture of these insects requires an ability to maneuver rapidly around stalks. Tenerals behave similarly on sedges at lake edges, but in those situations birds can move their heads and bodies rapidly from side to side because they are standing on the ground. Under these circumstances longer rather than shorter legs are advantageous. Brewer's are very effective in capturing tenerals among sedges at the edges of lakes.

Therefore, the best shape for foraging in grasses and sedge edges of marshes is poor for foraging on the vertical stalks of emergent vegetation and *vice versa*. Though the statistical results of laboratory foraging experiments did not show it, observations of the foraging birds indicated that the liabilities are worse for a longer-legged bird on vertical stalks than for a shorter-legged bird in sedges. Redwings and Yellowheads *do* capture insects at high rates along sedge edges of ponds. In fact, all foraging rate data presented in Chapter Four were gathered under those circumstances, because birds foraging on sedge edges are highly visible. Among cattails, however, birds are readily lost from view, and foraging sequences are hard to observe. Visibility is, moreover, best where sedges are shortest, that is, where they have been heavily grazed, and my observations were concentrated in those sites. The degree to which the edges of ponds in the arid and semiarid West were grazed by herds of bison is unknown, but large grazing mammals may well have enhanced foraging opportunities for blackbirds prior to the arrival of Europeans.

Competition in Trees and Bushes

Foraging in trees and bushes requires different techniques. Sagebrush branches are very dense, and large birds find it difficult to move through them. Among blackbirds, only female Redwings attempt to forage within sagebrush crowns, while Yellowheads, Brewer's and male Redwings, when they do forage in sagebrush, walk on top of the canopies and pick off insects visible from that position. Not surprisingly, it is Redwings that bring the largest numbers of lepidopteran larvae from sagebrush, because these insects cannot be captured by a bird foraging externally. Brewer's also bring fair numbers of larvae, probably capturing them by hopping up from the ground. Perhaps this is one of the reasons why Yellowheads, even though they forage extensively in uplands in the Potholes during the afternoons, seldom bring many terrestrial prey to their nests.

Redwings are also the only one of the three blackbirds that regularly forage in trees. The negative influence of trees near marshes on habitat selection by Yellowheads has already been documented in Chapter Three. Though Yellowheads are larger and stockier than Redwings, the difference is not great enough to account for their strong disinclination to forage in trees. Possibly their general distribution in treeless or nearly treeless regions has led to this behavior, despite a general morphology not too dissimilar from that of tanagers that do forage primarily in trees.

6.3. COMPETITION AND SIZE

The significance of size differences within and between species of blackbirds is difficult to determine. Intraspecific sexual size dimorphism is strongly associated with icterid breeding systems, polygynous species having males substantially larger than females, while in monogamous species size differences are much less (Orians, 1972; Selander, 1966). This pattern is readily interpretable in terms of the potential reproductive payoffs for males when successful competition gives access to multiple females rather than just one. This line of reasoning suggests that females of polygynous species are much closer to the optimum size for foraging and predator avoidance than are males, and that the larger males pay for their competitive success in direct behavioral encounters during the breeding season with a poorer ability to exploit the environment at some other critical time(s) of the year. The advantages of size in competitive encounters are clearly revealed by the dominance of Yellowheads over Redwings. The possible intraspecific significance of size differences for competition among Redwings is currently under investigation (Searcy, 1977).

Though large size does not make available any new prey items for blackbirds during the breeding season, it may do so in winter, when blackbirds are highly granivorous. Redwings should be able to husk most if not all seeds that Yellowheads can, but Yellowheads, because of their greater kicking and/or digging strength, might be able to uncover seeds unavailable to Redwings. Large size could also affect winter competition for seeds, but at that season blackbirds are not territorial and patterns of winter flock movement do not involve regular interindividual fighting for food items, so this is not likely. Yellowheads winter further south than Redwings and might encounter different size distributions of seeds on their wintering grounds, but there are no data on this point. Alternatively, larger size might be the result of competition with Redwings during the breeding season. The two species are completely sympatric over the breeding range of the Yellowhead and probably have been so for a long time. Therefore, Yellowheads always have to compete with Redwings for territories, but it is doubtful that such extensive size differences would have been favored exclusively by differential success in these encounters. In addition, this hypothesis cannot explain the larger sizes of female Yellowheads, because they are under no such comparable selective pressures during the breeding season.

Competition for territories does set limits to the number of male Redwings able to establish and hold territories in the areas in which I have worked. The great similarities in foraging modes and foods taken by Redwings and Yellowheads indicate that rankings of habitat quality should be the same for both. Therefore, territories taken by Yellowheads, which are invariably the better sites on the more productive lakes, would also be excellent territories for Redwings. Additional evidence for this interpretation comes from the fact that, as lakes have declined in quality at the Potholes due to carp invasions, the number of territorial male Redwings at first increases—concomitantly with a marked decline in the number of territorial male Yellowheads—before declining as the emergence of aquatic insects becomes almost totally eliminated by the fish. Data from Lyle Lake were given previously (Figure 5.1). A similar pattern also occurred on Herman Pond (Figure 6.3).

Therefore, it is likely that blackbirds, especially Redwings and Yellowheads, have mutually influenced one another's evolution. A reasonable view is that, because of the presence of Redwings, Yellowheads are somewhat larger than they would otherwise be and that they are restricted in their geographic range and the number of sites occupied within their breeding range. The breeding range of the Redwing apparently is not influenced by Yellowheads, but the number of sites occupied clearly is. In the

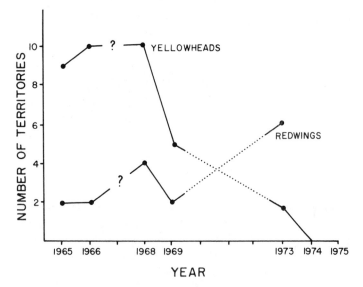

FIGURE 6.3. Annual changes in blackbird territories at Herman Pond, Columbia National Wildlife Refuge. The invasion of carp developed in 1968, causing great reductions in food availability and changes in blackbird populations.

absence of Yellowheads, which would make a large number of prime sites available to Redwings, selection would probably favor a slight size-increase in Redwing males if larger size were indeed advantageous in competition for sites where the number of females attracted per territory is unusually high. These are speculations, however, because the two species *have* evolved together. A detailed analysis of geographical variation in size of Redwings is currently being conducted by F. C. James, and her results should help to clarify the nature of selection for size.

These considerations do not, however, answer the question of why Yellowheads do not completely exclude Redwings from marshes in western North America. The proximate answer appears to be that Yellowheads are not sufficiently motivated or energetically able to establish and defend territories on areas of low productivity or where woody vegetation comes close to the shore. To explain this behavior pattern, however, it is necessary to determine the reasons why males that *do* elect to defend those territories, at which they would presumably be successful if they were sufficiently motivated, are less fit than males that do not attempt to do so. This is a particularly interesting point, because there are enough nonbreeding males for individuals rejecting poorer territories to stand a good chance of getting no territory at all. It must be true that males holding such territories have very low success in attracting females. This is equivalent to saying that there are insufficient females to make it advantageous for them to select low-quality territories under most circumstances. This, in turn, implies that the small number of females who survive the winter restricts the spread of Yellowheads into lower-quality marshes, but the factors governing overwinter survival are unknown for these and other birds. However, this argument does support Fretwell's (1972) contention that winter mortality rates strongly influence the structure of breeding bird communities. Unfortunately, this problem is an extremely difficult one to attack.

CHAPTER SEVEN

Adaptations Among Argentine Marsh-nesting Blackbirds

The theories developed and tested in previous chapters were molded by my experience with north temperate marshes and the birds breeding in them. Some of the theories were supposedly general, not being dependent upon any particular features of the birds or environments with which I was familiar, but theorizing is always influenced by empirical knowledge. An opportunity to test the generality of some of the theories was provided by the temperate marshes of South America, which have both closely related and distantly related species of breeding passerine birds. Therefore I measured prey availability in some Argentine marshes and studied foods and foraging, patch utilization, dietary breadths and overlaps of three species of south temperate blackbirds.

South temperate marshes differ from north temperate ones in a number of features. Probably most important is the fact that, since land masses in the southern hemisphere are comparatively small, climates at comparable latitudes are more maritime in the southern hemisphere than in the northern hemisphere. Also, temperate South America was not glaciated during the Pleistocene era, and the major marsh systems exist because of impeded drainage in very flat country rather than as a result of glacial activity as is often the case in North America and Eurasia. As a result, South American temperate marshes are for the most part accessible to, and contain substantial populations of, fish, whereas many northern potholes lack fish. Given the great impact fish may have on emergence of insects from marshes (Chapter Two), this is a difference of potentially great importance.

I made a short, preliminary visit to a number of marshes in eastern Argentina during December 1970. On the basis of information gathered at that time, I selected an area for more intensive study during the southern spring of 1973. An attempt was made to gather data similar to those obtained for northern blackbirds and northern marshes so that I could make independent tests of hypotheses developed in previous chapters of this book.

7.1. THE SPECIES OF BLACKBIRDS BREEDING IN ARGENTINE MARSHES

There are both tropical and temperate species among the South American marsh-nesting or marsh-associated blackbirds. The Chopi Blackbird (*Gnorimopsar chopi*) ranges from eastern and central Brazil south to the savannas along the Río Paraná in northern Argentina. The Unicolored Blackbird (*Agelaius cyanopus*) extends from just north of the Amazon River in northeastern Brazil, south along the Brazilian coast and inland in the drainage of the Río Paraná to the vicinity of Santa Fé, Argentina. The Chestnut-capped Blackbird (*Agelaius ruficapillus*) has an even more extended range from French Guiana south to Buenos Aires. In the southern, subtropical portions of their ranges these species are inhabitants of marshes of the Chaco, a region characterized by mild dry winters and hot rainy summers. Marshes in these areas are usually dry during winter but fill with water during the summer. As expected, the nesting periods of these species are during summer (December-March), just as Redwings in Costa Rica breed during the summer rains (June-September: Orians, 1973).

The temperate Scarlet-headed Blackbird (*Amblyramphus*

holosericeus) breeds sparingly in the Chaco of southern Brazil, Paraguay, Bolivia and Argentina but is more abundant as a breeding bird further south in temperate marshes of the Province of Buenos Aires. In these southern marshes it is joined by the smaller Yellow-winged Blackbird (*Agelaius thilius*), whose range extends south over most of Argentina to Santa Cruz wherever there are suitable marshes. It also breeds in high altitude marshes of the central and southern Andes from Peru southward. South of Buenos Aires Province it is the only marsh-breeding icterid in Argentina.

Other species forage at marsh edges and in moist pastures and regularly use marshes for nesting sites. The Yellow-rumped Marshbird (*Pseudoleistes guirahuro*) a species of restricted distribution in southern Brazil, Uruguay, eastern Paraguay and northeastern Argentina, is apparently nowhere common and is poorly known. The Brown-and-yellow Marshbird (*Pseudoleistes virescens*) has a more southerly distribution from extreme southern Brazil to the southern part of Buenos Aires Province, reaching its peak populations in temperate marshes in the eastern part of the latter. The Saffron-cowled Blackbird (*Xanthopsar flavus*) has a restricted distribution in coastal southeastern Brazil and Uruguay. It formerly bred sparingly in Argentina but has not been seen there since 1937 (Rumboll, pers. comm.)

My study areas were located near the coastal town of Pinamar, Buenos Aires Province, in the general region of Cabo San Antonio, where there are extensive marshes and lagoons (Vervoorst, 1967; Weller, 1967). The region has a mild temperate climate with a July mean temperature of 8°C and a January mean temperature of 20°C. Rainfall, which averages about 800 mm per year, falls fairly uniformly throughout the year, but because of high summer evaporation many shallower marshes dry up by March or April. The region was not glaciated, and marshes have apparently been formed by local tectonic movements and blockage of drainage systems by sand deposited by the strong northward-flowing offshore current (Cabrera, 1968; Vervoorst, 1967). Consequently, most marshes, even the large ones, are less than 2 meters deep, and stands of emergent vegetation cover most of many lakes.

The dominant emergent plant in most marshes is a bulrush (*Scirpus californicus*), and many large marshes are composed exclusively of this species (Figure 7.1). Immediately adjacent to the coast are extensive marshes of *Phragmites communis* and a large, coarse *Cyperus* (sp?). Cattails (*Typha dominguensis* and *T. latifolia*) are common in roadside ditches and also occur as almost pure patches in some of the marshes. Their habitat and successional relationships with *Scirpus* are not understood, but they probably invade

FIGURE 7.1. Typical marsh in the Pinamar area, Prov. Buenos Aires, Argentina. *Scirpus* dominates in the background, but there is a dense stand of a large, coarse *Cyperus* in the foreground. The open water is nearly covered by a floating mat of *Lemna* and *Azolla*. October 1973.

197

after disturbances and are replaced by *Scirpus* if disturbance does not continue. My evidence in support of this view is (a) Cattails dominate roadside ditches, which are highly disturbed habitats; (b) one area of roadside dominated by *Typha* in December 1970 was covered entirely by *Scirpus* in November 1973; (c) some marsh areas with cattails showed signs of recent disturbance such as grazing and burning; and (d) cattails were concentrated at the edges of marshes where grazing by cattle is very intense. Both cattail patches and bulrush beds are similar to comparable habitats in North America.

A marsh type without a North American equivalent is the undrained depression dominated by almost pure stands of the woody *Solanum malacoxylon*, which grows to heights of about 3 meters. All these depressions probably dry out completely during the summer.

My study sites were located adjacent to a 15 kilometer section of paved highway between Pinamar and General Madariaga. The highway offered all-weather access to marshes, and roadside ditches were favored nesting sites of blackbirds, especially early in the season (Figure 7.2). The area was worked daily from early October through November 1973. Observations were made on blackbirds in a variety of marshes but my efforts were concentrated in the following areas (names are mine):

Great Grebe Marsh—a deep-water, large marsh dominated by extensive stands of *Scirpus*, which covered most of the lake, The presence of many fish indicates a significant amount of permanent water. The highway crosses the deepest part of the lake.

Artiguet Marsh—another large, deep-water marsh connected by a narrow channel to the Great Grebe Marsh. Its emergent vegetation is almost exclusively *Scirpus*.

Screamer Marsh—a large, densely vegetated marsh with varied vegetation. Extensive stands of *Scirpus* are in-

FIGURE 7.2. Roadside vegetation between Pinamar and General Madariaga, Prov. Buenos Aires, Argentina. Clumps of Pampas Grass held nests of Yellow-winged Blackbirds and Brown-and-yellow Marshbirds. Foraging areas were provided by the sedges in the ditch next to the road and the wet pastures that can be seen in the distance. October 1973.

terspersed with patches of *Typha* and *Cyperus*. Where the road adjoins the marsh, *Typha* forms extensive beds.

Eucalyptus Marsh—a *Scirpus*-dominated shallow marsh connected to the Screamer Marsh at its southwest end. There were two patches of *Typha* and patches of *Solanum* mixed in with the *Scirpus* in the portion of the marsh I studied.

Harrier Marsh—an extensive shallow marsh adjacent to the coastal sand dunes. It contained the most extensive beds of *Typha* and *Cyperus* in the area. Water was everywhere less than 1 meter deep and levels fluctuated more widely, in response to local rainfall, than in the other marshes. It probably dries up entirely every summer.

199

Practically nothing has been published on the social organization or ecology of south temperate marsh-nesting blackbirds. Therefore, I summarize the results of my studies on their breeding ecology before considering resource availability, resource utilization, habitat selection, dietary breadths and competition. Data are necessarily less complete than for the North American species, because I gathered information during a single breeding season and worked simultaneously on three species. Nonetheless, the basic features of their breeding systems and resource utilization are clear from these studies.

Scarlet-headed Blackbird (*Amblyramphus holosericeus*). The Scarlet-headed Blackbird or Federal (Figure 7.3) was the least common of the three species of blackbirds breeding at Pinamar, and other investigators have also reported it to be present only as widely scattered pairs (Wetmore, 1926; Weller, 1967). The Federal is clearly a monogamous breeder. I was able to find nine nests, each of which was

FIGURE 7.3. Scarlet-headed Blackbird (*Amblyramphus holosericeus*) in bulrushes near its nest. Sexes are nearly identical. Pinamar, Argentina. October 1973.

200

attended by two adults. At every visit to an active nest I was scolded by two birds, and both are involved in all stages in the nesting cycle. I obtained no evidence that would suggest polygamy in this species.

My determination of the sexes of the nearly identical males and females was based on the following information: (a) in each pair one bird was slightly more brightly colored than the other, and in all cases this individual sang more than its duller mate; (b) the brighter bird of the pair was always more vigorous in pursuit of avian predators in the nest vicinity; (c) the brighter of the two birds did not incubate but usually perched conspicuously near the nest and often accompanied its mate on foraging trips; (d) the duller bird carried nest material much more often than the brighter one and was observed to solicit twice in the typical passerine female precopulatory posture. Therefore, I judge the slightly brighter, more vocal, predator-mobbing, nonincubating individual of each pair to be the male. Birds were either paired when I arrived in early October or they subsequently arrived already paired. Nonbreeding birds, especially first-year birds, were present in flocks around the marshes throughout the breeding season. I also observed postbreeding flocks of adults. Probably pair formation takes place early in spring in these flocks, but the actual process has not been observed.

After arrival at a breeding site the male spends most of his time perched conspicuously in the general area of the future nest. During this period he sings regularly, though not intensively by Redwing or Yellowhead standards. The female, on the other hand, spends most of her time low in the vegetation. She is almost impossible to observe but probably is exploring vegetation and food resources. A much smaller amount of time of actual foraging would suffice as males obtain their needed calories much more quickly.

A pair of Federales defends a large territory against all

other conspecific individuals. Both male and female partic-
ipate together in defense, much of which takes place in the
air and includes vigorous calling and a conspicuous
dangling-thighs display in which the red thigh feathers are
fully ruffled. Space defense behavior was directed against
arriving pairs of adults, apparently searching for a territory, and also against single individuals or groups of nonbreeding first-year birds. Pursuit of invaders continued for
long distances high in the air. A smaller area close to the
nest was defended against Yellow-winged Blackbirds, but
this was more sporadic, and I found nests of Yellow-
winged Blackbirds as close as 30 meters to an active Federal
nest. Wren-like Rushbirds (*Phleocryptes melanops*) were also
vigorously chased near nests.

Territories of Federales near Pinamar were very large.
The defended area of the Harrier Marsh pair extended
along the road for 750 meters and at least that far to the
north, constituting an area of about 560,000 square meters. Exact boundaries were difficult to determine on all
territories because in no case were territories of adjacent
pairs contiguous. Also, I never observed a territorial chal-
lenge at several of the nests and had to judge size strictly
from the area exploited when nestlings were being fed.
Adults regularly traveled at least 1 kilometer from the nest
to gather food for nestlings, and incubating females also
regularly flew comparable distances to gather their own
food. The effective territories of the nine pairs I studied
were all probably at least 40 hectares in extent and some
may have been larger. In some instances the entire territory lay within one contiguous large marsh, but some of the
territories included several smaller marshes and strips of
roadside ditches.

Federales preferred the sturdy stems of *Solanum* for
their nests, but seven of the nine nests were associated with
patches of *Typha* or *Cyperus* (Table 7.1). Nest building ap-

202

TABLE 7.1. Nest summary data, *Amblyramphus holosericeus*, Pinamar, Argentina, 1973.

Nest no.	Site	Date of first egg	Clutch size	Fate	Comments
2855	*Typha* patch, nest in *Solanum*	10/22	4	Adults prob. shot 11/10, young died	Marsh mostly *Scirpus* but with *Typha* patches
2849	Open, mixed *Typha* and *Scirpus*	10/24	3	Adults prob. shot 11/10, young died	1 egg disappeared during incubation
2899	Open roadside ditch, supported by 5 *Solanum* and 2 *Typha*	10/11	3	Fledged 2 young	Nest very conspicuous, all feeding areas along road
2875	*Cyperus* patch in *Scirpus-Typha* marsh	10/22	3	Fledged 3 young	—
2862	*Scirpus* marsh, supported by 6 *Solanum*	11/4	3	Prob. fledged 1 young	Nest open and conspicuous, 2 young disappeared 1st week
2891	Large *Cyperus* patch in *Scirpus* marsh	11/12	3	?	1.5 meters above water, attached to living and dead *Cyperus* stalks
2853	*Scirpus* area next to *Typha* patch, supported by 5 *Solanum*	?	3	?	Nest open and conspicuous
2857	Large *Cyperus* patch, attached to living *Cyperus*	11/22	3	?	General area mixed *Cyperus* and *Scirpus*
2903	*Scirpus* marsh with mixture of *Solanum*	?	3	?	Nest attached to *Solanum*

pears to be primarily but not exclusively performed by females. At the Great Grebe Marsh on October 12 the female made six visits to the nest with material between 0715 hours and 0830 hours while the male sat nearby and sang. At the Screamer Marsh on October 26 the female came in with material six times during the intervals from 0827 to 0935 hours and 1610 to 1815 hours. During the morning period the male came near the nest once with material but dropped it before going to the nest. At one nest on November 16 the female came in with material five times and the male three times between 0950 and 1035 hours.

Federal nests are almost identical to Yellowhead nests, being constructed entirely of strips of cattail leaves. The lining is more finely shredded strips of cattail leaves, and the entire nest is a woven cup supported by vertical stems. *Solanum* is probably the best nest site because it is the sturdiest marsh plant and also because there is no seasonal growth of the stems to which the nest is attached. All three nests in *Cyperus* were badly tilted by unequal growth of different supporting stems, and the eggs would almost certainly have fallen out of one had I not performed some selective adjustments on the attachments. Clutch size at Pinamar was generally three eggs (Table 7.1).

Rather surprisingly, I found no Shiny Cowbird (*Molothrus bonariensis*) eggs in any Federal nests. Whether they are not laid there or the Federal is a rejector (Rothstein, 1975) is not known, but nests of other blackbirds were very heavily parasitized. No Federal nests were taken by predators during the observation period, but at two nests both adults disappeared on October 11 and the tiny young starved. Probably all four were shot by one of the many people who slowly drove along the road on weekends shooting at any moving target. Given the heavy predation rates on nests of the other two blackbirds at Pinamar, the total absence of predation on Federal nests is surprising.

During the egg-laying period a male spends most of his time near the nest, whereas the female is usually absent and foraging. Incubation is performed exclusively by the female, and I never observed a male feeding an incubating female. The female leaves for brief periods to feed, normally unaccompanied by the male, who spends most of his time perched conspicuously near the nest while she is gone.

Once the eggs hatch, both birds spend the bulk of their time gathering food, but there is still a difference in sexual roles. Nests with young are seldom left unattended. Typically, one bird remains at or near the nest until the other returns with food. When nestlings are small the female usually spends this time brooding, but I never noticed a male brood nestlings. Instead a male sits near his nest on a conspicuous perch. Males and females feed at equal rates, and visits to the nest with food were made on the average once every five minutes (Table 7.2).

Yellow-winged Blackbird (*Agelaius thilius*). The widespread Yellow-wing (Figure 7.4) is nowhere as common as the Redwing in North America. Like it, however, it travels in large flocks during the nonbreeding season (Hudson, 1923; Weller, 1967). By the time I began my work at Pinamar, these flocks had largely broken up, but I observed small flocks regularly throughout the study period. First-year males were prominent in these flocks, but they also contained birds with fully adult plumage. Birds in flocks fed in wet pastures and at marsh edges.

My evidence indicates that the Yellow-wing is monoga-

TABLE 7.2. Rates of feeding of nestling *Amblyramphus holosericeus*, Pinamar, Argentina, October-November 1973. (y = young; d = days since hatching.)

Date	Nest	Age of young	Time	Visits by ♂	Visits by ♀	Comments
11/7	2855	2y-1d	1545-1715	8	12	Female brooded 9 times for a total of 28 minutes. Male sat near nest 8 times for a total of 35 minutes.
11/13	2875	3y-5d	0923-1030	7	6	
11/14	2899	2y-4d	1115-1215	3	6	
11/15	2899	2y-5d	0720-0850	6	5	Female brooded 5 times for a total of 43 minutes.
11/17	2899	2y-7d	0740-0914	11	11	
11/18	2875	3y-10d	0658-0828	10	14	No brooding: one adult around all the time except for 14 minutes. Male present for 66 minutes, female for 16.
				45[a]	54[a]	

[a] Average of one visit every five minutes.

FIGURE 7.4. The female Yellow-winged Blackbird (*Agelaius thilius*) is very similar to a female Redwing but is much smaller. Pinamar, Argentina. October 1973.

mous though it is strongly sexually dichromatic. At all three nests at which both adults were color-banded only those two adults visited the nest. In addition, I observed twelve other nests closely enough to be quite certain that one pair of birds was in attendance at each. I observed four sites where there were two or three nests within 10 meters of one another especially carefully as, if there were polygyny, one male should be mated to females of adjacent nests. In all four cases, however, each nest had a different male associated with it.

Pair formation apparently takes place prior to nest site selection. Many birds were already paired when I began work on October 10, and no breeding activity was initiated by unpaired birds. Typically, birds prospected for nesting areas together, the female leading as the pair arrived. The female immediately dropped down into the vegetation while the male perched high and sang. The female also initiated departure, the male quickly following. I never actually observed a pair selecting a site in which they subsequently built, but the behavior of prospecting birds was consistent.

The Yellow-wing at Pinamar was nonterritorial. At the site where four nests were within 10 meters of one another, I never saw site-related defense between pairs. At another nest observed extensively when young were being fed, I often saw prospecting birds fly into the clump of cattails containing the nest and actually visit the nest or sing within a few meters of it without evoking any antagonistic responses from either member of the pair, both of which were color-banded. I observed much fighting among males over females, but never over space.

Nests were placed in *Typha* and Pampas Grass (*Cordateria selloana*) clumps in roadside ditches, in tall grass in ditches, and in patches of *Typha* or *Cyperus* clumps in *Scirpus* marshes. Nests were close to the water and, in the case of *Typha* nests, in previous year's growth. Of 28 nests I found, 14 were in *Typha*, 8 in *Cyperus*, 3 in Pampas Grass and 3 in short grass. I found no nests in *Scirpus*, despite the fact that this was the most abundant plant in the general breeding habitat, nor any nest in *Solanum*. Because *Scirpus* is regularly used by both Redwings and Yellowheads in North America, its avoidance by *all* the blackbirds at Pinamar was unexpected.

Nest construction was primarily, if not exclusively, performed by females. I saw males with nesting material in their bills a number of times, but usually they dropped it soon after they picked it up. Only once did I actually observe a male go to a nest under construction with material in his bill and fly away from the site without it. Even in that case I am not certain that he actually worked on the nest rather than dropping his material close to it, because he was hidden in dense vegetation.

During 282 minutes of observations at nests with full clutches of eggs, I never observed a male visit a nest either to incubate or feed a female. Sometimes the male accompanied his mate as she left the nest to feed, but more often the two individuals fed at different times and in different places. Males often sat in the general nest vicinity and sang while females incubated. Of nineteen nests for which I was able to determine clutch size, there were 4 clutches of two eggs, 11 clutches of three, and 4 clutches of four eggs ($\bar{X} = 3.0 \pm 0.42$). The incubation period, at the two nests at which I was able to determine it, was thirteen days.

There was no sharp peak in nest initiation in 1973. Of those nests for which I was able to determine starting dates, four were started between October 10 and 20, three between October 21 and 31, five between November 1 and 11, two between November 11 and 20 and none between November 21 and 28. The decline in nest building may not have been as great as suggested, because *thilius* nests were very difficult to find (for me at least), and I did not locate many of them until they had young. Therefore I had a lower probability of finding nests started during the last week of field work.

Brown-and-yellow Marshbird (*Pseudoleistes virescens*). Outside the breeding season Brown-and-yellow Marshbirds are highly gregarious (Weller, 1967; Hudson, 1923), and large flocks of presumed nonbreeders are occasionally encountered during the breeding season. Details of the unusual social system of this species have been presented elsewhere (Orians, Orians and Orians, 1977) and will only be summarized here. The Brown-and-yellow (Figure 7.5) is the first blackbird to breed at Pinamar (egg laying began about September 20 in 1973). At this time the birds are monogamous, the male and female traveling together while feeding and during visits to the nest site. Nest building is done entirely by the female (detailed observations at

FIGURE 7.5. A color-banded Brown-and-yellow Marshbird (*Pseudoleistes virescens*) on a fencepost near its nest. This bird was known to be a female but the sexes are identical. Pinamar, Argentina. October 1973.

three nests). Males accompany females and occasionally pick up nesting material but never take it to the nest. Only the female incubates, but the male may feed her on the nest and, in the majority of nests, more than one bird brings food to the incubating female. The average clutch size was 3.78 eggs, and there was no evidence that more than one female laid in a nest. The number of birds attending nests further increases once the eggs hatch, a maximum of eight birds being observed bringing food to a single nest. The sexes, ages and relationships of these extra individuals are, unfortunately, unknown.

We found no evidence of territoriality in Brown-and-yellow Marshbirds. Nests were often situated in close proximity to one another, the shortest internest distance being 4 meters. Such grouping was usually associated with clumps of suitable pampas grass, the favored early spring nesting site when other herbaceous vegetation was still short and afforded little protection. Other nests were isolated more than 500 meters from their nearest neighbors. We observed only one case of intraspecific aggression near a nest. Adults from other nests often perched within a few meters of nests without being chased.

7.2. FOODS AND FORAGING

Because of the importance of aquatic insect emergence patterns for marsh-nesting blackbirds in western North America, I attempted to determine insect emergence patterns in marshes around Pinamar. Emergence cages were designed to be similar to those used in Washington, and I placed them in comparable habitat types. During October I built and installed a series of emergence cages in four study marshes. They were constructed of a heavy wire mesh, covered on top and all sides by plastic screening of 2 mm mesh (Figure 7.6). Each trap covered an area of 51 cm × 89 cm and was 25 cm high. Traps were visited in late afternoon on alternate days and insects counted and removed. I had intended to run all traps until the end of the study, but theft terminated sampling early on Great Grebe Marsh and the Eucalyptus Marsh roadside ditch. Dates of operations, number of traps and their locations are given in Table 7.3.

A second food resource for blackbirds is insects within stalks of emergent vegetation. These were sampled by splitting stems of *Scirpus* for 1 meter above the water surface. Insects were also taken from *Typha* and *Cyperus* by pulling apart the sheathing leaf bases. The final resource sampled was arthropods in the floating mats of *Lemna* and *Azolla* that covered the water in most marshes. This vegetation was sorted manually, a technique that did not extract all arthropods but which did provide a picture of kinds of prey present, their relative abundances and their sizes.

FIGURE 7.6. A floating emergence cage in a *Scirpus* marsh near Pinamar, Argentina. Notice the cut bulrush stems inside the trap and the dense floating mat of *Lemna* and *Azolla* on the water.

Compared with temperate marshes of western North America, the most striking feature of Pinamar marshes was the paucity of emergence of aquatic insects. Only four Ephemeroptera and one Trichoptera were captured during 1,243 trap-days. Diptera, usually numerically very abundant in north temperate marshes, also emerged in very small numbers, less than one individual per trap-day (Table 7.4). Numbers of emerging dragonflies (Anisoptera) at Pinamar were similar to values obtained in Washington (maximum 0.08 per trap per day), but emergence rates of damselflies (Zygoptera) were approximately 2 percent of that on good lakes in Washington (maximum 0.8 per trap/day compared to about 30/trap/day with higher values during peak periods). Emergences of odonates were better on larger, deeper lakes (Great Grebe, Artiguet) than on shallower lakes, but there was no discernible "edge effect" as in Washington. Traps placed on outer edges of

TABLE 7.3. Locations, numbers and dates of operation of emergence traps at Pinamar, Argentina, 1973.

Marsh	Description	Dates of operation	No. of traps	Total no. of trap-days
Great Grebe	Outer edge of deep *Scirpus* marsh; fish	Oct. 20-Nov. 14	3	75
Artiguet, shore	Shoreline of deep *Scirpus* marsh; fish	Oct. 31-Nov. 28	5	140
Artiguet, *Scirpus*	*Scirpus* beds in 1 meter of water near shore; fish	Oct. 31-Nov. 28	5	140
Eucalyptus, ditch	Open *Typha* in shallow ditch, no floating mat; no fish	Oct. 19-Nov. 24	2	72
Eucalyptus, *Scirpus*	*Scirpus* beds in ½-1 meter of water far from shore; some fish	Oct. 19-Nov. 28	8	320
Eucalyptus, *Typha*	*Typha* patch in large expanse of *Scirpus*, far from shore in ½ meter of water; some fish	Oct. 19-Nov. 28	7	280
Harrier, open	Open, grazed stand of *Scirpus* and *Typha*, in 0.2-0.5 meters of water, no fish	Oct. 23-Nov. 28	3	108
Harrier, *Cyperus*	Low, lightly grazed stand of *Cyperus* in 0.2-0.5 meters of water; no fish	Oct. 23-Nov. 28	3	108
		Total trap-days		1,243

beds of emergent vegetation did not capture more emerging insects than traps located within such beds. Also there was no difference between traps located in beds of *Scirpus* and beds of *Typha*. At the Artiguet Marsh, however, traps located on the shore of the lake caught 3-4 times the number of odonates as traps located out in the *Scirpus*.

There are several possible causes for poor rates of emergence of aquatic insects from Pinamar marshes. One is that the reduced seasonality, which allows avian predators to exploit larval insects the entire year, depresses the number successfully able to overwinter. A second possibility is that

TABLE 7.4. Numbers of emerging aquatic insects captured in traps at Pinamar, Argentina, October-November 1973.

Site	Anisoptera		Zygoptera		Diptera		Other		All insects	
	total	$\bar{X}$/trap/day	total	$\bar{X}$/trap/day	total	$\bar{X}$/trap/day	total	$\bar{X}$/trap/day	total	$\bar{X}$/trap/day
Great Grebe	6	0.08	30	0.40	29	0.39	1	0.01	66	0.88
Artiguet, shore	6	0.04	112	0.80	17	0.12	1	0.01	136	0.97
Artiguet, *Scirpus*	2	0.01	28	0.20	112	0.80	3	0.02	145	1.04
Eucalyptus, ditch	1	0.01	6	0.08	19	0.26	0	-	26	0.36
Eucalyptus, *Scirpus*	1	0.01	30	0.09	12	0.04	1	0.01	44	0.14
Eucalyptus, *Typha*	0	-	35	0.13	27	0.10	0	-	62	0.22
Harrier, open	4	0.04	6	0.06	3	0.03	0	-	13	0.12
Harrier, *Cyperus*	0	-	18	0.17	2	0.02	0	-	20	0.19
Total	20		265		221		6		512	

the dense floating mats of *Azolla* and other plants that characterize most marshes around Pinamar intercept most of the sunlight, greatly reducing photosynthesis by aquatic algae. As odonates are high on aquatic food chains they should accordingly have low population densities compared to those in lakes without extensive mats of floating vegetation. One of the sites, Eucalyptus Ditch, lacked a floating mat of vegetation, but it was shallow and dried up during the summer. It had few odonates, mostly from the family Lestidae, damselflies typical of temporary lakes (Walker, 1953). Greater penetration of sunlight could be responsible for higher emergence at the edge of Artiguet Marsh, where a peripheral strip about 10 meters wide, which was heavily grazed by cattle, lacked the floating mat. The fact that traps situated offshore from the grazed periphery caught so few odonates may indicate that individuals captured in shore traps were derived primarily from the grazed strip along the marsh edge rather than moving shoreward from deeper water. Many emerging insects in Washington marshes come from deeper water because shore traps exposed to the open water capture many more insects than shore traps behind beds of emergent vegetation (Chapter Two above). Third, the abundance of fish in most marshes may have reduced otherwise greater emergences.

Potential blackbird prey captured within stalks of emergent vegetation are shown in Table 7.5. *Scirpus* stalks opened randomly yielded almost no prey. Even when I selectively opened stalks with signs of possible internal infestation, I seldom found anything. The two lepidopteran larvae I discovered in the random sample of 100 stalks were much smaller than any individuals brought to nests by adult blackbirds.

Samples of the floating mat of *Lemna* and *Azolla* were sorted manually on October 15, November 6 and Novem-

TABLE 7.5. Animals collected within stalks of emergent vegetation at Pinamar, Argentina, October-November 1973.

Prey Taxon	Cyperus (100 stalks)	Dead Typha (200 stalks)	Live Typha (400 stalks)	Scirpus (100 stalks)
Dermaptera	5	31	19	
Hemiptera, Reduviidae			1	
Diptera, pupae		2		
Coleoptera				
Staphylinidae		4	5	
Hydrophyllidae		1	1	
Cerambycidae			1	
Chrysomelidae			4	
Curculionidae	1			
Helodidae		13		
? adults	2	9	10	
? larvae	1	26	16	
Lepidoptera		1	1	2
Insecta, pupa		1		
Araneae	16	17	21	
Isopoda		1	18	
Solpugida			1	
Acari		1		
Gastropoda	2			
Hirudinea	1			
Amphibia, Hylidae	2			
Total	30	107	98	2

ber 27. The fauna was dominated by talitrid amphipods of the genus *Hyalella*. My collection of 387 of these crustaceans was only part of the sample total. Other important prey items included gastropods (37), leeches (16), small fish (4), odonate larvae (4), belostomatid Hemiptera (15), Homoptera (4), Neuroptera (5), Ephemeroptera (4), Diptera (1) and Coleoptera (Hydrophyllidae, 11; Curculionidae, 2). Any bird able to probe through the mat, as would a rail, or able to manipulate the mat to expose the animals within it, has access to a rich food source. It would be difficult, however, for an adult blackbird gathering food for nestlings to capture multiple-prey items per trip, be-

cause each capture of a prey probably requires opening the bill widely. Therefore, this source, like arthropods in sheathing leaf bases, should be most valuable as food for adults or for species with precocial young.

Compared to passerines breeding in western North American marshes, those breeding at Pinamar exploit food supplies that cannot be captured at high rates and that are not suitable for delivery in large batches. This may explain why blackbirds were much less abundant in the Pinamar marshes than in North America. I had to devote much more time to find comparable numbers of nests, and the same was reported by Weller (1967) for Cabo San Antonio, just north of Pinamar.

7.3. PATCH UTILIZATION WHILE FORAGING

Emergence is in general very low at Pinamar, and areas of high emergence are either rare or absent, but the extent of my sampling is not great enough to exclude the possibility of occasional local concentrations of emerging insects. It is difficult to compare these results with data obtained by stem splitting or sampling the floating mat of vegetation from the point of view of probable capture rates by foraging blackbirds, but it is evident that a bird foraging by stem splitting or in the floating mat cannot under any circumstances capture prey at the high rates possible for blackbirds capturing emerging aquatic insects in Washington. In addition, bringing more than one prey per visit to a nest would be difficult. For these reasons patch selection and utilization should be very different for Argentine blackbirds than for North American ones.

Patch Use by the Scarlet-headed Blackbird

Most food for nestling Federales is gathered at considerable distances from the nest, and flights between 0.5 and 1 kilometer are normal. I could detect no tendency for one sex to forage closer to the nest than the other. Part of the reason for traveling long distances for foraging is that birds fly directly to patches of *Typha*, which are highly scattered among extensive beds of *Scirpus*. At nest 2855 for example, four small patches of *Typha* at distances from a few hundred meters to over a kilometer from the nest were the nearly exclusive sources of food for nestlings.

Despite much effort I seldom was able to see foraging birds low in dense vegetation. The few times they were in view they searched carefully among stems of emergent vegetation, selected a site, hammered into a stem and then split it with a powerful gaping movement of the bill. Arthropods within the stems were then extracted. At other times, especially when birds were foraging in patches of *Typha*, I could hear sounds of ripping vegetation but could not actually see the birds, because when I approached they always stopped feeding and moved further on.

Patches of *Typha* and *Cyperus* are probably selected for foraging because stems of *Typha* are a much richer source of invertebrates than stems of *Scirpus* (Table 7.5). The sheathing leaf bases of *Typha* and *Cyperus* afford excellent hiding places for a variety of arthropods, whereas *Scirpus* has no such sheaths. Nevertheless, I did see Federales splitting stems of *Scirpus* and extracting insects from them. Evidently, *Scirpus* stems occasionally contain larger prey than I found. Prey delivered to nestlings were mostly different from those I extracted from *Typha* stems (Table 7.5, Appendix N). For example, earwigs were prominent in cattail stems, but only one was recovered from food samples. Also, lepidopteran and coleopteran larvae in food samples were not the same species I found when splitting stems. Evidently the birds, at least when gathering food for nestlings, were foraging in somewhat different sites from those I sampled.

My observations of birds scolding me at nests suggest that Federales bring a single food item to a nest per trip. Since nests were visited once every five minutes on the average, this amounts to only twelve prey items per hour per nest, in other words, usually four per nestling. While foraging, adults regularly changed locations and often departed from the nest in a different direction than the one they came from. Theoretically, birds should confine their foraging activities to the best patches and not leave them until the encounter rate with prey has been reduced to the average for all patches in the optimal set. If this is the case, birds should shift their foraging locations after trips during which it took longer than average to find the single prey item. Therefore, I compared foraging trips when a bird departed from the nest in a direction different from the one from which it had arrived with those in which arrival and departure directions were the same. Birds changed directions 11 times when the previous search time was longer than the one before it but 8 times when the foraging trip was actually shorter in duration than the one before it. Considering all foraging trips, the mean gathering time when birds departed from the nest in the same direction as they had arrived was 4.24 ± 3.29 minutes while the mean gathering time was 6.44 ± 5.21 minutes when birds left the nest in a different direction from that from which they had arrived. The difference is not statistically significant because variances are large and birds regularly changed foraging directions after trips of very short duration. These results are difficult to interpret because exact foraging locations are unknown and because birds may or may not have fed themselves before gathering food.

Nineteen food samples covering 20 hours of delivery of prey were obtained from three Federal nests (Appendix N). In addition I was able to identify some insects carried by adults scolding me at the nests. They included four brown pupae or larvae, one teneral damselfly and one teneral dragonfly, complete with wings.

Only 7 percent (7 of 99) prey delivered to nestling Federales were insects with aquatic larval stages, but as I never saw a Federal foraging except in a marsh with at least shallow standing water, presumably all prey were associated with aquatic vegetation. The scarcity of emerging insects in the food samples is not surprising given the small amount of emergence actually taking place. The food samples also support my direct observations that the majority of prey taken are not visible on leaf or stem surfaces but must be revealed by gaping.

Patch Use by the Yellow-winged Blackbird

During the breeding season at Pinamar, Yellow-wings foraged in a wide variety of habitats, mostly with standing water. A favored site was wet areas in pastures after heavy rains. Birds foraged mostly at the edges of standing water where prey were captured by direct pecking movements, but they also gaped into grasses protruding above the water. Another favored foraging site was old *Typha* stalks where birds gaped into the leaf sheaths to expose hidden prey. I also observed birds gaping into plant material clinging to the bases of cattail and bulrush stalks. While perched on stalks of emergent vegetation, birds also regularly gaped into the mat of floating *Azolla* or grabbed bits of the mat in their bills and overturned plants with a rapid sideways movement of the head. The Yellow-wing is a small blackbird and moves easily among stalks of emergent vegetation. It forages readily from vertical stalks and regularly perches between stalks with legs straddling to opposite stems. From my observations it appears that, during the breeding season at least, they obtain the bulk of their food by gaping into vegetation.

Both male and female Yellow-wings regularly bring food

to nestlings. During 330 minutes of observations at nests with young I counted 20 visits by males (once every 16.5 minutes) and 52 visits by females (once every 6.4 minutes). Between visits females virtually never stayed around their nests, but males regularly did so, often singing from a conspicuous perch though not defending any area around the nest. Nevertheless, the difference in visitation rate cannot be accounted for by the time males sat near the nests. Though I was unable to observe birds well enough while they were foraging to know whether males had less success in finding prey, it is highly unlikely that females can capture prey more than twice as fast as males. Probably males engaged in other activities, such as advertising and courtship while away from their nest.

Usually, but not always, a single prey was delivered per trip. On November 5 at nest 2828, which contained three young, two days old at the time, the female brought teneral damselflies on six of her eighteen visits to the nest. On five of those visits she definitely had but a single individual, but on one visit she had at least two and possibly more. At the same nest on November 9 the female also scolded me with a single teneral damselfly in her bill. At a marsh 16 km west of Pinamar I observed a male with two teneral damselflies in his bill on November 12. At the same marsh I also observed a female with several small prey in her bill, probably midges. On another occasion I saw a male with two brown pupae in his bill. It is difficult to judge the percentage of visits to the nest with more than one prey because even when I was watching at close range birds often came to the nest so rapidly or at an angle such that I was unable to count the prey in the bill.

Nineteen food samples, representing 20 hours of delivery of prey were obtained from nestling Yellow-wings during November (Appendix N). Of the three species of marsh-nesting blackbirds at Pinamar, the Yellow-wing is the only one that extensively uses emerging aquatic insects.

Sixty-two percent (69 of 112) of prey delivered to nestlings were of aquatic groups and odonates alone accounted for 49.1 percent of prey in the food samples. Of the 55 odonates in the samples, 28 were teneral Lestidae, damselflies characteristic of ponds that dry out in summer. The others were mostly larvae and tenerals of coenagrionid damselflies and aeshnid and libellulid dragonflies more typical of permanent waters. These data suggest that *thilius* regularly forages in shallow ephemeral ponds. These ponds did provide the few cases I observed of significant emergences of damselflies at Pinamar. Both food samples and direct observations indicate that *thilius* is a generalized aquatic forager that captures emerging aquatic insects when they are available, gapes in the floating mat of vegetation, and splits stems of cattails.

Patch Use by the Brown-and-yellow Marshbird

Brown-and-yellow Marshbirds foraged primarily in wet pastures and marsh edges. Most prey were captured by digging into the turf and bases of clumps of grasses with their long, strong bills. None of the prey they captured are visible to a human observer peering into the vegetation even at close range. Thirty-nine food samples covering 46 hours of delivery were obtained at nests of Brown-and-yellows. Only 3 out of 195 prey items (2 percent) taken in the food samples were of taxa with an aquatic stage some time during the life cycle (Appendix N). Rather, samples were dominated by lepidopteran larvae (50.5 percent) and spiders (29.9 percent). I was unable to devise a practical way of determining prey abundances in the wet pastures where birds foraged. In fact, even by careful digging in the turf in places where I had watched birds foraging with considerable success, I was seldom able to find any of the prey they typically brought to the nests. Therefore I cannot evaluate selection of foraging patch or prey items in this species.

Selection of Prey Items

Capture rates of large prey by any of the three black-birds never approached that required for them to reject small prey. Nevertheless, as each prey item requires a round trip to the nest, prey size dominates the determination of delivery rate of energy. Therefore, it would pay to continue searching for large items to deliver to the nest while consuming small prey (Orians and Pearson, 1979). This would be especially true for Federales, which forage at such great distances from their nests. Unfortunately, I was not able to observe adults foraging for nestlings regularly enough to be able to tell whether they do eat smaller prey, selectively bringing only larger prey to nestlings. Also, I lack quantitative estimates of prey sizes in the habitats in which birds were foraging to be able to compare them with prey actually delivered to the nestlings, as I did for Redwings in Costa Rica (Orians, 1973).

7.4. DIETARY BREADTHS

Dietary breadths for the three species of blackbirds, calculated by the index $H = 1/\Sigma p_i^2$, where p_is are proportions of prey in the categories in Appendix N, are given in Table 7.6. *Agelaius thilius*, a generalized forager for aquatic insects, has the broadest diet, while the stem-splitting *Amblyramphus* has the narrowest. Breadth is similar to that of Redwings and Yellowheads (Table 5.17) in Washington.

7.5. COMPETITION AND OVERLAP

Overlap in prey delivered to nestlings at Pinamar, calculated from the data in Appendix N using Pianka's (1973) index of overlap, is presented in Table 7.7. Overlap in taxa of prey is low among all three species. Overlap in foraging patches is also low, and when the species do forage in the

TABLE 7.6. Dietary niche breadths ($H = 1/\Sigma p_i^2$) for Argentine marsh-nesting blackbirds, measured by the prey delivered to nestlings.

Species	Breadth of prey delivered to nestlings
Pseudoleistes virescens	4.421
Agelaius thilius	6.010
Amblyramphus holosericeus	7.241

same places, they use very different foraging modes. In patches of emergent vegetation *thilius* searches for superficially available insects and works over the mat of floating vegetation, while *Amblyramphus* hammers into and splits stems. In wet pastures *Pseudoleistes* captures nearly all of its prey by digging into the turf, while *thilius* feeds superficially. *Amblyramphus* and *Pseudoleistes* seldom forage in the same habitats.

Therefore, in all aspects of their foraging ecologies as revealed by direct observation and prey delivered to nestlings, the three Argentine blackbirds are less similar to one another than the three marsh blackbirds in the Pacific Northwest. Why they should be so is less clear. One might argue that in Washington food is so abundant during the breeding season that high overlap is not disadvantageous, but this argument is weakened by the existence of nestling starvation in my study areas. Alternatively, the massive emergence of aquatic insects in Washington causes odo-

TABLE 7.7. Overlap in prey delivered to nestling blackbirds, Pinamar, Argentina, as measured by Pianka's (1973) index of overlap, α_{ij}.

	Pseudoleistes virescens	Agelaius thilius	Amblyramphus holosericeus
Pseudoleistes virescens	1.0	.152	.188
Agelaius thilius		1.0	.255
Amblyramphus holosericeus			1.0

nates and dipterans to be very common in both aquatic and upland patch types so that taxonomic overlap of prey is inevitably high. In addition, the abundance of prey available superficially may have acted as a selective force opposing the evolution of specialized stem-splitting behavior and morphology in north temperate blackbirds. It could also be true that winter rather than breeding season overlap is more critical for Argentine blackbirds than for North American blackbirds.

The striking differences in availability of food during the breeding season between marshes in Argentina and the Pacific Northwest have probably had a major effect on the evolution of ecology and social organization of blackbirds in the two continents. High rates of emergence and great temporal and spatial patchiness of that emergence in North America have led to high densities of breeding blackbirds and conditions that favor polygynous mating systems. Under the variable conditions that formerly prevailed in lowland California, this led to the evolution of extreme coloniality in Tricolored Blackbirds (Orians, 1961). There is evidently insufficient spatial patchiness in productivity for female blackbirds to benefit by selecting an already mated male in Argentina, and all three marsh-associated blackbirds are monogamous. Only one of the three species is territorial, and it (*Amblyramphus*) defends very large territories. The other two seem to be somewhat opportunistic in where they forage, but I know too little about the temporal and spatial patchiness of their prey to be able to evaluate that aspect of their foraging or to suggest why their nests show the clumping patterns they do.

The fact that most prey delivered to Argentine nestlings must be extracted from hiding places by pecking and/or gaping movements makes it difficult for foraging adults to bring more than one prey per trip to the nest, as also occurs for the same reason among Redwings in Costa Rica

(Orians, 1973). This, combined with lower availability of prey in general, is probably the reason for the small clutches of Argentine blackbirds despite the fact that adults of both sexes, augmented by helpers in *Pseudoleistes*, feed nestlings in all three species, while of the North American species, both sexes feed regularly or at a high rate in only the Brewer's Blackbird, which has the largest clutches of the three species.

These great differences in marsh characteristics, most of which were not expected prior to the initiation of my sampling program in Argentina, made it difficult to test some important aspects of the theories developed in North America. Bypassing small prey is not to be expected in Argentina, and there is no time of day or place where encounter rates with large prey are likely to be high. Habitat patchiness is not sufficient to favor polygyny, and I lack sufficient information about patchiness to test theories about settling patterns of breeding birds. However, given the low availability of resources in Argentine marshes, the patterns of ecology and social organization among Argentine blackbirds are to be expected, and my data—though incomplete—add useful information about the probable influence of productivity on avian social ecology.

CHAPTER EIGHT

Of Birds and Marshes

As a result of many years of study I tend to view marshes as factories producing food and providing nest sites for blackbirds. This bias has dominated previous chapters of this book and the research upon which they are based, but marshes are utilized by other kinds of birds and animals from other taxa. Deeper insights into the evolution of blackbird adaptations are attainable by considering them in a somewhat broader context, including the effects of the birds upon their prey.

Freshwater marshes have a number of features of major significance to the birds that exploit them. Their vegetation is structurally simple and often dominated, sometimes completely, by a single species of emergent plant. The species and genera involved are widespread, and some are found in marshes on all continents. Food resources, at least of some types, have great temporal and spatial variability in most marshes, the significance of which differs for most passerines vs. nonpasserines. Finally, marshes constitute a small fraction of the total landscape in most areas. Usually they are "islands" in a sea of upland habitats, and the lack of a solid substrate sets marshes apart from terrestrial habitats with respect to foraging opportunities and patterns of prey availability.

8.1. SIGNIFICANCE OF MARSH STRUCTURAL SIMPLICITY

As a result of the work of Robert H. MacArthur and studies stimulated by him, it is well known that the physical structure of vegetation, especially the vertical distribution of foliage (foliage height diversity), is strongly correlated with the number of species of birds breeding in a habitat (MacArthur and MacArthur, 1961; MacArthur, MacArthur and Preer, 1962; Cody, 1966, 1968, 1974; Karr, 1971; MacArthur, Recher and Cody, 1966). This correlation is not perfect, and the causes for it are as yet incompletely understood, but an obvious inference is that the number of distinct competitively viable modes of foraging in any habitat is strongly correlated with vertical complexity of the vegetation. Supporting evidence comes from studies of bird communities in vegetation which, though structurally complex, is nonetheless relatively uniform vertically. For example, pine forests are dominated by trees that have needle densities and characteristics that do not change much with height. In addition, because the trees are open canopied multilayers (Horn, 1971) much of the light entering a canopy penetrates to the ground. Therefore, birds foraging low in branches encounter relatively bright substrates almost identical to those which would be encountered high in the canopy. Significantly, species of birds foraging in pine foliage utilize nearly the full vertical range of foliage in both temperate and tropical pine forests (Balda, 1969).

Structurally simple habitats, such as marshes and grasslands, differ from forests in that the available foraging space is highly compressed vertically, but within that short distance conditions for foraging may change rapidly. Light interception is often nearly complete and the bases of plants provide strikingly different foraging opportunities from those of their upper parts. In grasslands, the best known of presumed structurally simple habitats (Cody, 1966, 1968, 1974; Wiens, 1973, 1974), the number of breeding bird species does not change much with latitude. Freshwater marshes differ from grasslands in several important respects that affect breeding birds. First, marshes lack a solid soil substrate. This eliminates or makes certain foraging modes much more difficult and, in addition, deprives arthropods of a class of hiding places that are extensively used in grasslands. Because of the absence of terrestrial hiding places at the bases of marsh plants, places within and on stems of emergent plants are especially important for arthropods unable to move below the water surface for more than a short period of time. Therefore, structural features of different species of marsh plants should have a more important influence than structural features of grasses on the number of breeding bird species.

Because of the wide ranges of the common genera of emergent marsh plants, marshes can be compared intercontinentally in terms of their dominant plants. A convenient classification is into (a) Bulrush (*Scirpus*) marshes; (b) Cattail (*Typha*) and Sedge (*Cyperus*) marshes; (c) Reed (*Phragmites*) marshes; (d) Mixed marshes with grasses, sedges, cattails and bulrushes; and (e) Wet meadows, dominated by grasses, rushes and sedges. In addition, the unusually tall papyrus (*Cyperus papyrus*) marshes of Africa and the Middle East are sufficiently different from all the above to contain breeding birds rare or absent in other marsh types. Foraging opportunities available in each of these marsh types are shown in Table 8.1. The table simplifies a complex resource picture, but the differences listed among marsh types are widespread and generally important to passerine birds. For example, cattails and bulrushes do have seeds but they are so small that they are not utilized by any species of bird as a food source. Similarly, cattail and bulrush marshes may have substantial floating mats of vegetation, in which case the mat is available for exploitation, but these mats tend to become more prevalent as a marsh fills in, which is when grasses and sedges normally become mixed with the cattails and bulrushes. Reed (*Phragmites*) beds are of interest because they combine structural fea-

TABLE 8.1. Foraging opportunities for passerines provided by different types of marshes.

Marsh type	Foraging opportunities
Pure bulrush (*Scirpus*)	Arthropods on surfaces of stems (few) Arthropods within stems (small and few) Arthropods at water surface
Pure cattail (*Typha*)	Arthropods within stems (few) Arthropods on surface of stems Arthropods within sheathing leaf bases Arthropods at water surface
Reeds (*Phragmites*)	Same as *Typha* plus seeds
Papyrus (*Cyperus papyrus*)	Same as *Typha* but taller and denser

Mixed bulrush, cattail, sedges and grasses	Arthropods on surfaces of stems
	  (variable stem types)
	Arthropods within stems
	Arthropods within sheathing leaf bases
	Arthropods hidden in bases of fine clumps
	  of grasses and sedges
	Arthropods in surface mat of vegetation
	Seeds

Sedge-rush-grass meadows	All of the above plus: Arthropods hidden in wet ground

tures of a cattail marsh with availability of seeds large enough for birds to exploit. In many areas of the world, reed beds support a few species of birds absent or rare in the other types of marshes.

Table 8.2 lists the species of passerines that typically breed in different marsh types in temperate North America and Argentina (for scientific names of these birds see Appendixes O and P). On both continents pure *Scirpus* marshes are the poorest in species, and as Redwings and Yellowheads are interspecifically territorial, no spot in North America has more than two of the three species listed for *Scirpus* marshes. The three species characteristic of this habitat in eastern Argentina are widespread in temperate South America, and also form the basic passerine bird community in *Scirpus* marshes of the high *altiplano* (above 3,000 meters) of Bolivia and Peru (pers. obs.).

TABLE 8.2. Passerines typically found breeding in different types of marshes in central North America and Buenos Aires Province, Argentina.

Marsh type	Breeding passerines	
	North America	Argentina
Pure bulrush	Red-winged Blackbird Yellow-headed Blackbird Long-billed Marsh Wren	Yellow-winged Blackbird Wren-like Rushbird Many-colored Rush-tyrant
Pure cattail	Yellow-headed Blackbird Red-winged Blackbird Long-billed Marsh Wren Song Sparrow	Scarlet-headed Blackbird Yellow-winged Blackbird Wren-like Rushbird Many-colored Rush-tyrant Curve-billed Reedhaunter Sulphur-bearded Spinetail Warbling Doradito
Cattails, sedges	Yellow-headed Blackbird Red-winged Blackbird Long-billed Marsh Wren Song Sparrow (Swamp Sparrow in East) Common Grackle [a] Brewer's Blackbird [a]	Yellow-winged Blackbird Scarlet-headed Blackbird Wren-like Rushbird Many-colored Rush-tyrant Curve-billed Reedhaunter Sulphur-bearded Spinetail Warbling Doradito Long-tailed Reed-finch Rufous-collared Sparrow Great Pampa-finch
Sedge-rush-grass meadows	(Manitoba) Red-winged Blackbird Bobolink Western Meadowlark Brown-headed Cowbird Short-billed Marsh Wren Long-billed Marsh Wren Northern Yellowthroat Song Sparrow Savannah Sparrow Clay-colored Sparrow	Yellow-winged Blackbird Brown-and-yellow Marshbird Correndera Pipit Spectacled Tyrant Black-and-white Monjita Bay-capped Wren-spinetail Grassland Yellow-finch Great Pampa-finch
Reed marshes		Spectacled Tyrant Warbling Doradito Curve-billed Reedhaunter Sulphur-bearded Spinetail Yellow-winged Blackbird

[a] Feed in marshes but usually nest elsewhere.

As the richness of foraging opportunities increases from *Scirpus* marshes to complex wet meadows, so does the number of breeding passerines in both continents, but richness is generally greater in Argentina than in the United States. Most marsh passerines in Argentina are flycatchers (Tyrannidae) and ovenbirds (Furnariidae). In North America there are flycatchers that breed on marsh borders but none are typical of nonshrubby marshes. Ovenbirds are an ecologically diverse group that utilizes substrates and food sources exploited in North America primarily by wrens and sparrows. Exact equivalents are generally not discernible, but the Wren-like Rushbird is remarkably similar in appearance, behavior and breeding biology to the Long-billed Marsh Wren.

A pure bulrush marsh provides the smallest number of foraging substrates and has the smallest number of breeding bird species in both continents. The emergent plants have no sheathing leaf bases and few insects are present superficially on stems. In each continent only one species uses this substrate, the Long-billed Marsh Wren in North America and the Wren-like Rushbird in Argentina. The third Argentine species, the Many-colored Rush Tyrant, is a tiny flycatcher that gleans for small insects at the bases of the stems and from the floating mats of *Lemna* and *Azolla*.

Cattails have a more complex structure, and their sheathing leaf bases provide hiding places for a variety of arthropods. The greater variety of foraging modes of passerines in cattails compared to bulrushes is shown in Table 8.3. In this environment the disparity between North American and Argentine marshes in bird species richness appears to be the greatest.

In shallower water, where grasses and sedges are mixed with cattails and bulrushes, several new resources become available. There is a significant seed source, there are clumps of fine grasses into which foraging birds can probe,

TABLE 8.3. Utilization of cattail (*Typha*) marshes by breeding passerine birds in North America and Argentina.

	Species	
Method of utilization	North America	Argentina
Pecking and stem splitting	—	Scarlet-headed Blackbird
Gaping into leaf sheaths	Yellow-headed Blackbird Red-winged Blackbird	Yellow-winged Blackbird
Probing into leaf sheaths	Long-billed Marsh Wren	Wren-like Rushbird Curve-billed Reedhaunter Sulphur-bearded Spinetail
Surface gleaning	Yellow-headed Blackbird Red-winged Blackbird Long-billed Marsh Wren Song (or Swamp) Sparrow	Yellow-winged Blackbird Wren-like Rushbird Curve-billed Reedhaunter Sulphur-bearded Spinetail Many-colored Rush-tyrant
Snatching from surfaces	—	Warbling Doradito

and the accumulation of debris provides a substrate upon which small birds can walk. Not surprisingly the number of species of breeding passerines is higher in this habitat than in the previous ones. This same trend continues into wet sedge, rush, and grass meadows where, interestingly, richness in North America, at least in the highly productive northern prairie meadows, exceeds that in Argentina. In this environment a new method of foraging is utilized by passerines dropping to the ground from an elevated perch. This is the usual mode of hunting for two flycatchers (*Hymenops perspicillata* and *Xolmis dominicana*) at Pinamar, Argentina. In addition, probing into the mat of vegetation in shallow water is now possible, and *Pseudoleistes virescens* and *Sturnella neglecta* do this in Argentina and North America respectively. Seeds are produced in greater quantities, and more species of sparrows and finches breed here than in deeper water marshes. It is not certain that this relationship is a causal one, however, since most sparrows feed primarily on insects during the breeding season.

Reasons for greater avian species richness in Argentine

232

marshes are not evident. One possibility is that temperate marshes of South America are generally found at low latitudes with maritime climates. At Pinamar where I worked winters are mild, the marshes never freeze, most passerine birds are resident, and even migrants travel relatively short distances to southern Brazil and Paraguay. Many southern flycatchers are morphologically similar to tropical species rather than north temperate species, that is, they have large heads and short tails, possibly because they are not adapted for long-distance migrations. Contrary to this interpretation, however, the cattail marshes I studied in Costa Rica did not have any breeding flycatchers (except on the edges, as in North America), and the breeding bird community was extremely simple, consisting of Red-winged Blackbirds, Groove-billed Anis (*Crotophaga sulcirostris*) and White-collared Seedeaters (*Sporophila torquata*). Though in the tropics, these marshes varied seasonally, completely drying out during the long dry season, filling rapidly with water when the rains began, and often fluctuating rapidly according to frequency and severity of rains.

An alternative interpretation employs concepts from the theory of island biogeography. The temperate marshes of Argentina are large and almost continuously distributed from the delta of the Paraná south along the Atlantic coast for several hundred miles. To the north they are continuous with vast subtropical marshes of the Paraná Valley, extending uninterrupted northward into Paraguay. Most of the more southerly breeders are also found in this area, though some of them are much less common there than further south. In contrast, temperate marshes of North America are much more patchily distributed over a wide area and only in the far north, where summers are short and migration distances are long, are marshes large and extensive. The number of species of organisms that evolve

adaptations to a particular environment is strongly correlated with its extent (Terborgh, 1973). The smaller the
area of a habitat, the smaller a selective force it exerts on
organisms, most of whose dispersing propagules fall on
other sites, though this argument is less persuasive for
birds which are highly dispersive and have active habitat
selection. Small, local populations are probably more
prone to extinction, and if their habitat type differs much
from the predominant ones, dispersal from other habitats
may be low.

It is more difficult to compare breeding bird communities between eastern and western hemispheres because of the paucity of data from Eurasia, Africa and Australia. Also lacking are data on quantities and temporal
patterns of emergences of aquatic insects from marshes
in those areas. Preliminary data from Switzerland suggest
very modest emergences more comparable to those found
in Argentina than those of western North America (R.
Furrer, pers. comm.).

Valuable comparative data on structure of marsh passerine communities are provided by the work of Zahavi
(1957) in Israel and Ruwet (1965) in Katanga, Zaire. In
both areas a range of marsh types was studied (Table 8.4).
Papyrus was absent in the Katanga study area, though it is
a dominant plant in many East African marshes (Lind and
Morrison, 1974). In Israel the marsh passerine community
is composed almost exclusively of warblers, and no finches
are reported to breed in any of the marsh types. Most
marsh types have two species of warblers, one of which
forages high and one of which forages low in the vegetation (Zahavi, 1957). In Katanga, in addition to warblers, a
number of species of weaverbirds (Ploceidae) are characteristic of marshes and a cuckoo (*Centropus monachus*) is
widespread in beds of *Cyperus* and *Typha*.

In East Africa, papyrus beds are reported to contain

TABLE 8.4. Breeding passerines of marshes in Israel and Katanga, Zaire (data from Zahavi (1957) and Ruwet (1965).

	Breeding passerines	
Marsh type	Israel	Katanga
Papyrus (*Cyperus papyrus*)	*Acrocephalus stentoreus*	
	Cettia cetti	
Phragmites beds	*A. arundinaceus*	*Textor melanocephalus*
	A. scirpaceus	*Hyphanturgus ocularis*
	Locustella luscinioides	
Typha-Cyperus beds	*A. scirpaceus*	*Calamocichla gracilirostris*
	A. melanopogon	*Estrilda subflava*
		E. astrild
		Vidua macroura
Wet, grassy meadows	*Cisticola lincidis*	*Cisticola galactotes*
	Motacilla flava	*Saxicola torquata*

three species of reed warblers (*Acrocephalus*), a fan-tailed warbler (*Cisticola carruthersi*) and a shrike (*Laniarius nfumbiri*) (Moreau, 1966), but no detailed studies of marsh bird communities appear to have been made there. West African marshes also contain several species of warblers (*Acrocephalus baeticus*, *Bradypterus brachypterus*, *Calamocetor leptorhyncha*) and weaverbirds (*Euplectes afra* and *Coliuspasser hartlaubi*) (Bannerman, 1953), but information on their exact breeding habitats is unavailable.

The same groups of species extend across Asia to Australia, where all marsh types appear to have one or more warblers, some small finches and often a cuckoo. For example, in Burma, reed beds contain *Acrocephalus arundinaceus* (Great Reed Warbler), a weaverbird (*Ploceus manyar*) and an estrildine finch (*Lonchura ferruginosa*). Grassy marshes and meadows have a small thrush that drops on prey from an elevated perch (*Saxicola leucura*), a warbler (*Megalurus palustris*), a weaverbird (*Ploceus manyar*) and two estrildine finches (*Lonchura ferruginosa* and *Estrilda amandavi*) (Smythies, 1953).

In northwestern China, Wilder and Hubbard (1938) report reed beds to contain two species of reedbuntings (*Em-*

beriza yessoensis, *E. pallasi*), two titmice (*Remiz pendulina*, *Panurus biarmicus*) and three species of *Acrocephalus*. In the vicinity of Sydney, Australia, reed beds harbor two sylviid warblers (*Acrocephalus australis*, *Megalurus gramineus*) and the Chestnut-breasted Finch (*Lonchura castaneothorax*) (Hindwood and McGill, 1958).

Passerine community structure in Old World marshes differs strikingly from that in the New World. First, there appear to be no equivalents to icterids in any Old World marshes. The relatively large Great Reed Warblers approach the size of small female blackbirds, but they are reported to forage mostly high in tall marsh vegetation and probably are not dependent on emerging aquatic insects to the extent that blackbirds are. The Great Reed Warbler appears to be adversely affected by cold, rainy weather during the breeding season, and there is a negative correlation between brood size and fledging weight in that species in Poland but not in Reed Warblers breeding in the same marshes (Dyrcz, 1974). This suggests that the larger Great Reed Warbler may be dependent at least partially on emerging aquatic insects, a view supported by their favoring edges of reed beds rather than extensive closed areas of reeds (Dyrcz, 1974).

Phragmites beds are much less extensive in North and South America than they are in Eurasia, and there appear to be no species entirely or primarily restricted to this habitat in the New World. The most prominent Eurasian species nearly confined to reeds, the Bearded Tit (*Panurus biarmicus*), extensively utilizes the small seeds of *Phragmites*. These seeds *are* utilized in the New World, but the species that do so are not restricted to that habitat, probably because of its relatively small extent compared to the extensive Eurasian reed beds.

Unexpectedly, no members of the large and ecologically varied starling family (Sturnidae) have evolved into marsh breeders. These birds are similar in size to blackbirds and also have well-developed muscles for forcibly opening their bills (Lorenz, 1949). Since I believe that the ability to gape was very important for the evolution of icterids and their success in exploiting marshy habitats, this failure is especially puzzling. Some species feed at the edges of ponds and marshes in a number of areas, but true marsh breeding is not reported for any starling. The lack of evident ecological counterparts to icterids in Eurasia may help explain the conclusion of Corbet (1962, p. 116) that Zygoptera are not heavily preyed upon by birds during emergence.

Though it may be a result of the paucity of data, there is no convincing evidence that bird species richness increases in marshes affording more varied foraging opportunities in the Old World. However, I have been able to construct Old World marsh bird communities only for Israel and Katanga.

Throughout the world in tropical and subtropical regions, cuckoos (nonpasserines) are a common component of land bird communities in marshes. This niche is filled primarily by anis (*Crotophaga*) in the New World and coucals (*Centropus*) in the Old World. In the New World, anis extend to latitudes at which winters become cold enough for orthopterans to tend to overwinter as diapausing eggs rather than as adults. Orthopterans are major components of anis' diets at all times of the year, as illustrated by Groove-billed Anis in northwest Costa Rica (Table 8.5), and are probably especially important during the tropical dry season when lepidopteran larvae are unavailable.

Another potentially significant consequence of the structural simplicity of marshes is the restricted nature of nest sites available for birds. Marsh vegetation provides little protection from strong winds or heavy rain, and it is relatively easy for predators to search for nests in the low vege-

237

TABLE 8.5. Prey taken from stomachs of adult Groove-billed Anis (*Crotophaga sulcirostris*) at Taboga, Guanacaste, Costa Rica, July-August 1967.

Order	Family	No.	Prop. of total
Odonata	Coenagrionidae	1	.003
	Libellulidae	1	.003
Orthoptera	Acrididae	40	.122
	Blattidae	1	.003
	Tettigoniidae	7	.021
	Tetrigidae	1	.003
Dermaptera		3	.009
Homoptera	Cercopidae	4	.012
	Cicadellidae	22	.067
	Membracidae	1	.003
	?	1	.003
Hemiptera	Coreidae	4	.012
	Pentatomidae	30	.091
	Pyrrhocoridae	6	.018
	Reduviidae	1	.003
	Scutelleridae	18	.055
	?	10	.030
Lepidoptera	Noctuidae	81	.246
	?	51	.155
Coleoptera	Carabidae	11	.033
	Chrysomelidae	14	.043
	Curculionidae	9	.027
	Lampyridae	1	.003
	Tenebrionidae	1	.003
	?	2	.006
Araneida	Lycosidae	3	.009
	?	4	.012
Acarina	Ixodidae	1	.003
	Total	329	

tation. Counteracting this to some extent, mammals must swim in all but the shallower waters, and many terrestrial nest robbers do not usually enter marshes. Conceivably, the correlation observed between complexity of hiding places in marsh vegetation and the number of breeding passerine bird species might be explained, entirely or in part, by the richness of nest sites for the birds themselves. Nest predation rates are very high on marsh passerines (Holm, 1973; Orians, 1973; Robertson, 1972; 1973a, b; Willson, 1966), and nest site availability does influence the number of females attracted to territories of polygnous species (Holm, 1973; Verner and Engelsen, 1970; Zimmerman, 1966). Nevertheless, it is doubtful that nest site availability is an important factor influencing the number of *species* of breeding birds in marshes, because all species use essentially the same sites. Their nests, though of varied construction, are all attached to vertical stalks of emergent vegetation. No use as nest sites is made of the varied hiding places for prey identified in Table 8.1 most of which are too small to accommodate nests of even the smallest passerines. Individual species may favor denser patches of emergent vegetation (Redwings, Long-billed Marsh Wrens) or more open patches (Yellowheads), but a suitable nest site for any one of the species is suitable for the others. The situation is quite different, however, when the much larger nonpasserines are considered.

8.2. NONPASSERINE BIRDS IN MARSHES

Marshes are extensively exploited by nonpasserine birds throughout the world, and in some areas, especially high latitudes, nonpasserines dominate marsh bird communities. Nonpasserines differ from passerines in several ways that cause them to be influenced by different factors than those critical for passerines.

First, nonpasserines are generally larger than passerines, often considerably so. This restricts their foraging from stems of emergent vegetation. Only the smaller species, for example Least Bitterns, regularly do so. On the other hand their larger size increases the depth of water in which they can wade. Differences in length of legs and bills strongly influence water depths at which nonpasserines can feed (Kushlan, 1976). Long bills also permit probing into substrates to extract food not available to surface feeders.

Second, many nonpasserines have nidifugous young capable of following their parents around soon after hatching. Whether or not these young feed themselves, adults do not have to deliver food to a central place, which increases flexibility in nest placement relative to feeding areas. In fact, many species such as ducks and rails regularly place their nests long distances from feeding areas to which they then journey only once with their newly hatched young.

Even for nonpasserines with nidicolous young, which must be fed for long periods of time in the nest, their larger size makes it possible to return with relatively larger loads of food and, hence, to exploit food sources long distances from the nests. Wood Storks (*Mycteria americana*), for example, may forage as far as 130 kilometers from their nests (Kahl, 1964), and many sea birds travel much further than that. By comparison, the distances traveled by marsh passerines are very short, even for such densely colonial species as the Tricolored Blackbird (Orians, 1961).

The large size of nonpasserines makes their nests very conspicuous, and escape from predators by concealment is unlikely except for rather rare species. This gives high value to nesting in unusually safe sites, favoring colonial nesting. Most nidicolous wading and littoral birds nest in colonies in safe sites while nearly all nidifugous species nest solitarily with concealed nests (Lack, 1968).

Most nonpasserines exploit prey in the water or on the surface of floating vegetation, and they divide resources on the basis of size, water depth and depth in the substrate. Therefore, nonpasserines are affected differently from passerines by resource availability. For marsh passerines, depth of water is important primarily in influencing probabilities of nest predation while for nonpasserines it may determine whether a bird can forage or not. Feeding patterns of large wading birds in marshes are strongly determined by changes in water levels, which both affect potential foraging sites and also concentrate prey which are trapped in ponds that may dry up in the dry season (Kushlan, 1976; Mock, 1975). In fact, large wading birds are much more abundant and richer in species in semi-arid regions or ones having long and severe dry seasons. Herons, storks and ibises of wet tropical regions are rare, solitary foragers, while most species of drier regions are common, conspicuous and colonial.

Ducks and geese are also more abundant in marshes of drier regions, but their response to productivity differences is probably very similar to that of passerines. Aquatic insects exploited by most breeding ducks and grebes are most abundant in marshes having the greatest insect emergences. The management of marshes for waterfowl usually enhances blackbird populations, which is one of the reasons that I have found it profitable to study blackbirds on wildlife refuges.

8.3. AVIAN SOCIAL SYSTEMS IN MARSHES

Polygyny is unusually prevalent among passerines nesting in marshes (Verner and Willson, 1966; Orians, 1972; Verner, 1964). Blackbirds contribute strongly to this trend, but polygyny is also known among members of the Troglodytidae (Verner, 1964, 1965), Sylviidae (Catchpole, 1974) Paridae (Burckhardt, 1948) and Emberizinae (Bell and Hornby, 1969) breeding in marshes. Moreover, the icterids of western North America are exceptional in their strongly developed polygyny. This appears to be related to the unusual productivity of western North American marshes though adequate emergence data are lacking for most regions of the world.

Perhaps more interesting is the rarity of polygyny among nonpasserines in marshes (Lack, 1968). The large size and conspicuousness of these species probably results in prohibitively high nest-predation rates unless one adult is in attendance throughout incubation and until nestlings are large enough to defend themselves from attack by predators and conspecifics. Only rare, solitary species can afford to leave eggs and small young unattended. Furthermore, females of large species are unlikely to benefit from mating with a male already having a female unless there is an absolute shortage of males. Male parental investment is too valuable to favor accepting conditions of mandatory sharing.

The large size and mobility of nonpasserines, however, strongly favors coloniality. The great temporal and spatial variations in food availability in marshes, combined with the value of unusually safe nest sites, leads to selection of safe sites advantageously situated with respect to foraging areas (Horn, 1968). Coloniality is especially prevalent among species with nidicolous young having a long period of dependency in the nest (Lack, 1968). Among species with mobile young, nesting dispersion depends primarily on whether or not the young are safer in the feeding areas than are nests and eggs. If young attended by adults can feed relatively safely in areas where nests would be highly vulnerable, selection favors dispersed nests but clumped feeding groups. This pattern is especially common among ducks (Lack, 1968), where the young are highly mobile and can escape rapidly to cover when a predator is discovered.

8.4. ISLAND BIOGEOGRAPHY OF MARSHES

Though the number of species of birds breeding in marshes is probably about what one would expect from the vertical structural simplicity of marsh vegetation, it is reasonable to consider whether the isolated, patchy nature of marshes might have restricted the evolution of marsh-inhabiting species. The highly variable resource base in marshes could cause a high rate of local extinction. In addition, if living in marshes requires different adaptations from living in terrestrial habitats, dispersing individuals of many species would find it difficult to find sufficient food outside marshes. If marshes were highly scattered, this would reduce the probability of finding other marshes and should favor reduced tendencies to disperse overland.

On the other hand, the high productivity of marshes, by increasing the range and types of resources present above threshold levels, could function to increase the number of species living together. This appears to be unlikely, however, because of the great diurnal, seasonal and annual fluctuations of resources. In fact, the adaptations of many marsh birds appear to be molded strongly by resource fluctuations, as I have demonstrated in some detail for blackbirds. Many larger wading birds depend on water fluctuations, which trap prey in areas where they are especially vulnerable (Kushlan, 1976; pers. obs. in the Argentine Chaco).

For several reasons, the isolation of marshes is likely to be less significant for birds than the isolation of true islands. First, passerine birds living in marshes appear to have no physiological or morphological peculiarities. In fact, A. A. Allen long ago puzzled about the success of Redwings in marshes when he could detect no obvious adaptations to that environment (Allen, 1914). Marsh-breeding blackbirds are not obviously different physiologically and morphologically from upland-breeding ones, and the same is true for other marsh-breeding passerines. Indeed, because of the seasonal nature of food resources for passerines in marshes, many species utilize upland environments for much of the year, entering marshes only to breed. Redwings have even become one of the most abundant breeding birds in hayfields, croplands, pastures and abandoned fields over much of eastern North America, in-

243

dicating that their traits are well suited to a variety of different habitats, many of them without water and emerging aquatic insects. Though breeding success is sometimes lower in upland habitats (Case and Hewitt, 1963; Robertson, 1972), in some areas upland-breeding birds do as well as their marsh-breeding counterparts (Dolbeer, 1976; Robertson, 1973a, b).

Second, the patchy nature of marshes favors highly motile adults in a number of organisms that live in marshes. Most marsh-living insects have wide distributions and rapidly colonize new ponds and marshes. For example, though most Potholes lakes are less than thirty years old, and some of them considerably younger than that, they have the full complement of odonates expected of lakes in that region. Even such weakly flying birds as rails readily find isolated marshes, and most of them are able to survive in grassy upland habitats if they do not immediately find a marsh during their migrations.

For these reasons it is unlikely that the number of species of breeding birds in marshes, except in unusual circumstances, is substantially affected by the isolated nature of marshes. Resource availability and predation pressures are apparently much more powerful molders of marsh bird community structure and the attributes of individual species.

8.5. EFFECTS OF BLACKBIRD PREDATION ON ODONATE POPULATIONS

The focus of this monograph has been on effects of resource availability in space and time on breeding attributes of marsh-nesting blackbirds. It is also of interest to ask what are the effects of the birds on their prey? My study was not directed toward answering this question, but many of the data gathered are pertinent to it.

244

Predation by birds might affect insect populations in two major ways. First, it might be severe enough to influence overall population densities by restricting the number of egg-laying adults. Second, the birds might act as selective agents influencing the life-history characteristics of the insects; for example, temporal patterns of emergence. These effects could be independent of one another, because the level of predation required to affect population densities is much higher than that required to mold phenotypic traits.

Blackbirds and Odonate Population Densities

To estimate probable effects of blackbirds on insect population densities one must know insect emergence rates, blackbird predation rates, and how many emerging insects are necessary to provide enough eggs to saturate the capacity of a lake to produce insect larvae. It is difficult to make these estimates, particularly the last one, because I have not studied larval ecology of aquatic insects, and data available in the literature are few (Gambles, 1963; Kormondy and Gower, 1965; Macan, 1964). Nevertheless, even crude estimates are of some interest.

At Coot Lake in the Potholes, emerging insects were sampled with five shore traps and, as there are no beds of emergent vegetation at the lake, these data can be used to estimate total emergences of odonates, the group most effectively captured by the traps (see Chapter Two). The five shore traps sampled approximately 3 meters of the 500-meter perimeter of Coot Lake (0.006 of the total). During the main emergence period (May-August) the traps captured 47 odonates, mostly damselflies, per meter of lake edge per day, or 23,500 per day for the entire lake. Over an emergence season of 125 days, this yields a total of 2,937,500 odonates.

When traps were being emptied every two hours to determine the diurnal pattern of emergence, I counted all foraging blackbirds at the lake from a bluff at its east end. These counts yielded an average of 5.9 birds per observation (see Figure 4.7). My extensive counts at Coot and other Potholes lakes show that a foraging blackbird captures about 14 insects per minute. Therefore, the blackbird population was capturing 82.6 insects per minute or 69,384 for a 14-hour foraging day. The proportion of these prey that were odonates can be estimated from the food sample data taken at the Potholes (Chapter Four; Orians and Horn, 1969). Using the prey habitat assignments of Orians and Horn, and pooling prey for the two aquatic habitat patches (emergent vegetation and sedge edge), odonates constituted 0.796 of total prey items taken by blackbirds. Therefore, during a 14-hour foraging day blackbirds may have taken 55,230 odonates from the shores of Coot Lake, over twice the estimated emergence. Using proportion of prey taken only from sedge edges, odonates constitute 0.695 of all prey, yielding a total of 48,222 odonates per day.

Errors exist in my estimates of total emergence and predation rates by the birds. Counts of birds foraging at the lake are accurate but slightly inflated by the fact that the counts span only 12 hours rather than 14 and the times at both ends of the day are ones when few blackbirds forage at the lake. A more reasonable estimate of total blackbird predation might be obtained by assuming an average of 5 blackbird-minutes per minute which would give a total predation of 58,800 prey per day, or 46,805 odonates, a figure still twice the calculated emergence.

Another potential source of error is in estimating percent of prey taken that were odonates. Data were pooled from blackbird nests at a number of lakes, and I cannot use data from Coot Lake because none of the foraging blackbirds, except for a few of the Brewer's, actually bred on Coot Lake. Nevertheless, direct observations of foraging birds at Coot Lake indicate that large prey, nearly all of which are odonates, constitute 40.3 percent of prey captured by birds foraging there. One can adjust the estimates by making less-likely assumptions than those employed above. For example, if we assume that all Diptera were taken at the edges of lakes, then odonates would constitute only 0.536 of all prey captured there, yielding an estimated daily capture of 37,190 odonates (using a figure of 5.9 birds) or 31,500 (using the figure of 5 birds). Alternatively, if we calculate total predation by using percentage of *all* prey delivered to nestling blackbirds at the Potholes that were odonates (0.444), total daily captures of odonates at Coot Lake would amount to 30,806 (26,093 using 5 birds). These estimates, all of which involve unlikely corrections, still yield predation totals greater than estimated emergences.

A more likely source of error is that the emergence traps underestimate emergences. To enter a trap, an insect must swim underneath a screen that can presumably be perceived directly and which does reduce light intensities. Perhaps a certain fraction of larvae might avoid trap sites and emerge adjacent to them, but there is no way at present to evaluate this potential source of error. To do so would require constructing some barrier that would exclude birds but would not affect emergence behavior of the insects. We do not have a suitable design for such a device. Counts of exuviae do not provide useful estimates because many of the emerging insects are captured while they are still larvae. In addition, many exuviae fall rapidly into the water where they are impossible to recover.

Finally, we used only a small number of traps which, when combined with the great site-to-site variation in emergence rates, makes it likely that different placement of the traps would have yielded substantially different estimates of total emergence. Nonetheless, the numbers of odonates captured in our traps are in accordance with our visual impressions of relative abundances at the different lakes, and the numbers reported here are very high compared to estimates of emergences elsewhere. It is evident that emerging odonate larvae readily entered our traps, but this does not exclude the possibility that some of them did not. However, even if we assume that half of the larvae that would have emerged at the trap sites avoid them, this adjustment would still suffice only to increase calculated emergences to the general level of calculated total blackbird predation. Thus, though there are several uncertainties, it is difficult to avoid the conclusion that blackbirds are taking a very high proportion of emerging insects there.

Indirect evidence suggests, however, that birds are ineffective in influencing total odonate population densities. First, if Coot Lake produces about three million odonates (mostly damselflies) every year, half of which are females, and if each female produces 1,500 eggs during her lifetime (Grieve, 1937), it would require only 2,000 females escaping predation during the emergence period to lay three million eggs. Clearly, not all individuals escaping blackbird predation during emergence survive to adulthood, and not all eggs yield emerging larvae. However, even if 90 percent of individuals die between egg laying and the time they become vulnerable to blackbird predation, only 20,000 egg-laying females would be required. As this is only 0.67 percent of estimated emergence, which obviously errs in being too low, it is clear that only a very small percentage of emerging individuals, perhaps of the order of 1-2 percent, must escape. Looked at in another way, blackbird predation would have to be at least 98-99 percent effective to affect population size if other sources of mortality account for no more than 90 percent of individuals between adulthood and emergence. This conclusion is further strengthened if egg and larval mortality has a significant density-dependent component.

The speed with which odonates colonize new lakes at the Potholes, and high emergence rates from lakes less than two decades old, also suggest that availability of suitable habitats for larvae is the prime determining factor of odonate abundances. Also, the data presented earlier for carp and noncarp-infested lakes indicate a major effect of fish predation on odonate emergences, and similar results have been reported elsewhere (Macan, 1966a, b). Therefore, despite apparently very heavy blackbird predation rates during emergence, odonate populations are probably relatively uninfluenced by this mortality source compared to events within the lakes. Other investigators (Corbet, 1962, p. 198; Kormondy and Gower, 1965) also concluded that larval adaptations are the key to odonate ecology.

Effects of Blackbirds on Phenotypic
Traits of Odonates

Though blackbird predation may not exert much influence on odonate population sizes, it is clearly a powerful selective agent on life-history characteristics of odonates. For example, the synchrony of emergence during late morning hours may represent a "predator saturation" pattern which would be favored by natural selection if an insect emerging in the main emergence period had a slightly better chance of escaping predation than one emerging earlier or later in the day. Clearly, predation is not the only factor affecting emergence times: time from emergence from the water to flying would be shorter later in the day than at the time of peak emergence because afternoon temperatures are higher than morning ones. However, wind velocities are in general much higher in the afternoon, and freshly metamorphosed tenerals are susceptible to wind damage to their wings and may be blown into the water or be damaged when they attempt their first flight (Pajunen, 1962). Thus, the late morning peak of emergence is probably a compromise between selection by predators for synchrony at a time when the period of vulnerability is shortest and selection for avoidance of emergence during the windiest parts of the day.

Night is the safest time for odonates to emerge, and nearly all species emerge at night in the tropics where nights are warm (Corbet, 1962). The percentage of species emerging at night decreases with increasing latitude, the smaller species generally switching to diurnal emergence at lower latitudes than the larger ones. In Washington all damselflies and libellulid dragonflies emerge during the day while only the very large Aeshnidae, Cordulegastridae and Gomphidae emerge at night. Even the large dragonflies, however, appear to require some warming of temperatures in the morning before they can fly and are, accordingly, vulnerable to blackbird predation for the first hours of the day. In good weather, individuals of most species are ready to fly within an hour after dawn, but maiden flights can be delayed until midday by cold weather (Robert, 1958). Most of the dragonflies taken by blackbirds in the Pacific Northwest are captured before 0900 hours. Birds foraging at the edges of lakes in the Potholes early in the morning are often obviously searching in the water for emerging libellulid naiads or for aeshnid tenerals not yet ready to fly. My tropical experience is limited to Costa Rica where newly emerged odonates were apparently able to fly early enough in the morning to avoid blackbird predation (Orians, 1973).

Increased predator saturation would be achieved if the seasonal period of emergence of odonates were shorter than it is. In Washington the common species of damselflies emerge over several months, and though each dragonfly species, as is typical of temperate latitudes (Lutz and

Pittman, 1970), has a short period of emergence (often less than a few weeks), the emergence times of the species are staggered. These patterns suggest that larval competition is a more powerful molder of season of emergence than predation during emergence, because such timing patterns mean that different species are less likely to have similar-sized larvae in the water at the same time. Details of emergence of odonates at the Potholes will be analyzed in detail with respect to species and sex of individuals elsewhere (Cook and Orians, in prep.).

That blackbirds may exert only modest effects on total population sizes of their prey, though apparently being strongly limited by their prey, appears to be due to the short period of prey vulnerability to blackbird predation, combined with high reproductive rates of prey that make it possible for a small percentage of individuals to produce enough eggs to cause larval competition in the water. For birds exploiting prey available to them for longer periods, perhaps throughout their life cycles, but especially during their main feeding periods (as is the case with orthopterans and most lepidopterans), the potential for population as well as phenotypic effects is greater. Consideration of the effects of predators on both population dynamics and phenotypic traits of prey under conditions of different patterns of predator and prey life cycles is an area in which a great deal of fruitful work remains to be done.

8.6. CONCLUSIONS

The structural simplicity of marsh vegetation appears to be the major factor restricting the number of species of passerines breeding at any site, and bird species may be strongly associated with particular dominant species of plants. Nonpasserines respond quite differently to marshes, being much more sensitive to changes in depth of water than structure of vegetation. The large size of most nonpasserines permits them to exploit parts of marshes where passerines cannot forage but, by virtue of their size, they cannot hide their nests and must, therefore, be monogamous. Coloniality is also favored by large size because areas far distant from nests can be exploited during the breeding season and because the vulnerability of their nests places a premium on safe sites that are normally patchily distributed. Though the density and success of breeding blackbirds are strongly affected by prey availability, blackbirds probably do not exert a comparable influence on their prey even though they evidently can take a very high proportion of emerging individuals. Prey abundances are probably determined largely by competition and predation in the water, but we are largely ignorant of those dynamics.

General Conclusions

Though many of the cues by which blackbirds assess and select breeding habitats are still unknown, they are apparently able to judge them well enough that females approximate an Ideal Free Distribution in the Pacific Northwest. Females choose nesting areas primarily on the basis of habitat features rather than the characteristics of the males occupying them. This conclusion is supported by the behavior of females while settling and the results of experiments in which males are removed. Given that traits of males are probably more difficult to judge than are those of the marshes, this is not a surprising result. Habitat choices of Yellowheads may be simpler than those of Redwings because the former are restricted to very high-productivity lakes and gather a higher proportion of their food closer to their nests.

Marsh-nesting blackbirds share with other birds an ability to judge qualities of foraging patches and to shift their foraging activity in accordance with diurnal and seasonal patterns of prey availability. They do more sampling of different patch types at times of day when patches are changing rapidly in quality and, hence, it is less evident where foraging is best. Redwings reject small prey in Costa Rica, where they bring single items per trip to the nest, but do not in the Pacific Northwest where they are multiple-prey loaders. These patterns are all predicted by foraging theory, and one may conclude that blackbirds are making choices by rules similar to those proposed theoretically even though rigorous tests were not possible in the field.

Argentine marsh-nesting blackbirds are different from their North American counterparts in social organization, foraging behavior and extent of overlap in their use of resources. These differences are reasonably interpreted as adaptations to the much lower availability of prey in the Argentine marshes I studied and the corresponding lack of major differences in quality of breeding sites within and between marshes. Therefore they provide some support for the generality of some of the general theories of habitat, mate and prey selection that guided the design of my field work. As I have mentioned previously, many theoretical predictions were not clearly testable, but my intellectual forays back and forth between data and theory contributed substantially to the measurements I regarded as important and to the types of interpretations I am able to offer for field data.

Many of the questions unasked and unanswered by this study can be investigated only by following the choices of individually marked birds within and between breeding seasons. Obtaining this type of information on patterns of individual choices has been the top priority in the research carried out during the time the results of this extensive study were being analyzed and put into words. This ordering of objectives of the project is appropriate, because the more detailed examination of individual choices can now be set in the context of the general ecological conditions most likely to influence payoffs from those choices.

Appendixes

Appendix A. Vegetation density measurements on blackbird territories, Turnbull and Columbia Wildlife Refuges, 1966, 1968. Xmax = 203 cm.

Lake	Year	N	$\overline{X}$	S^2	S.D.	S.E.	C.V.	Type of territory
Beaver Pond	1966	102	37.0	604.8	24.6	1.7	66.3	All RW
" "	1968	56	38.2	629.4	25.1	2.4	65.7	All RW
Blackhorse Lake	1966	31	38.6	399.2	20.0	2.5	51.7	All YH
" "	1968	26	38.3	397.5	19.9	2.8	52.1	Mostly YH
Little McDowell (South)	1966	83	24.8	410.6	20.3	1.6	81.6	Both spp.
" "	1966	71	30.7	505.9	22.5	1.9	73.3	Both spp.
Little McDowell (South)	1968	50	40.7	776.6	27.9	2.8	68.6	Both spp.
" "	1968	25	39.5	802.0	28.3	4.0	71.7	All RW
" "		25	42.5	834.9	28.9	4.1	68.0	All YH
Little McDowell (North)	1968	48	44.0	731.5	27.0	2.8	61.4	Mostly RW
Big McDowell	1968	20	40.4	586.7	24.2	3.8	60.0	All YH
Mann Lake	1966	77	40.9	913.3	30.2	2.4	73.8	All RW
" "	1968	70	28.7	542.2	23.3	2.0	81.0	All RW
Kepple Lake	1966	58	39.1	688.3	26.2	2.4	67.1	Both spp.
Lyle Lake	1966	65	26.0	122.0	11.0	1.0	42.3	All RW
" "	1968	50	17.6	114.5	10.7	1.1	60.7	All RW
Willow Pond	1966	50	39.1	729.4	27.0	2.7	69.0	All RW
North Hampton	1966	121	25.2	377.4	19.4	1.2	77.0	All YH
" "	1968	62	27.8	458.7	21.4	1.9	76.9	Mostly YH
Herman Pond	1966	70	26.4	231.1	15.2	1.3	57.5	Both spp.
" "	1966	17	21.0	93.3	9.6	1.6	45.7	All RW
" "	1966	53	28.2	264.2	16.3	1.6	57.6	All YH
" "	1968	60	20.6	159.5	12.6	1.2	61.3	Both spp.
" "	1968	32	22.1	121.2	11.0	1.4	49.8	All YH
" "	1968	28	18.7	201.1	14.2	1.9	75.7	All RW

255

APPENDIXES

Appendix B. Vegetation density measurements on blackbird territories, Cariboo Parklands, British Columbia, 1966. Xmax = 203 cm.

Lake	N	$\bar{X}$	S^2	S.D.	S.E.	C.V.	Type of territ
Sorenson Lake	100	68.0	368.7	19.2	1.4	28.2	Mostly YH
Near Phalarope	100	59.3	573.1	23.9	1.7	40.4	All YH
Rush Lake	144	80.0	-	-	-	-	All YH
Westwick Lake	88	69.6	338.0	18.4	1.4	26.4	All YH

256

APPENDIXES

Appendix C. Caloric values assigned to categories of prey of blackbirds, based on burning of the species most commonly encountered in the food samples. Groupings other than families are enclosed in parentheses.

Order	Family	Calories
Collembola		5
Ephemeroptera	Baetidae	18
Odonata	(Zygoptera)	50
	(Anisoptera)	400
Orthoptera	Tettigoniidae	300
	Gryllidae	400
	Acrididae	300
	Locustidae	150
	Tettigidae	150
Dermaptera		40
Homoptera	Cercopidae	50
	Cicadellidae	50
	Cicadidae	800
	Delphacidae	25
	Fulgoridae	50
	Aphididae	25
Heteroptera	Corixidae	100
	Nepidae	565
	Notonectidae	75
	Cydnidae	30
	Pentatomidae	50
	Lygaeidae	30
	Reduviidae	50
	Mesoviliidae	50
	Nabidae	30
	Miridae	30
	Gerridae	50
	Veliidae	30
	Saldidae	30
	Belostomatidae	50
	Rhopalidae	50
Neuroptera	Raphidiidae	20
Trichoptera	(small)	10
	(large)	100
Lepidoptera	Arctiidae	200
	Noctuidae	200
	Nymphalidae	100
	Geometridae	200
	Lycaenidae	100
	Zygaenidae	100
	Pyralidae	200
	Pterophoridae	25
	Gelechiidae	100
	Oecophoridae	100
	Tortricidae	200

APPENDIXES

Appendix C. (cont.)

Order	Family	Calories
Diptera	(Nematocera) (unid.)	15
	Tipulidae	50
	Culicidae	15
	Chironomidae	15
	Ceratopogonidae	25
	Simuliidae	15
	Mycetophilidae	15
	(Brachycera) (unid.)	15
	(Orthorrhapha) (unid.)	15
	Stratiomyidae	25
	Rhagionidae	15
	Tabanidae	100
	Asilidae	25
	Bombyliidae	25
	Dolichopodidae	15
	(Cyclorrhapha) (unid.)	15
	Syrphidae	25
	Muscidae	25
	Tachinidae	25
	Tetanoceridae	15
	Drosophilidae	10
	Ephydridae	15
Coleoptera	Carabidae	150
	Cicindellidae	100
	Dytiscidae (?)	150
	Dytiscidae, large	370
	Dytiscidae, medium	150
	Dytiscidae, small	30
	Haliplidae	50
	Hydrophilidae	150
	Staphylinidae	75
	Cantharidae	100
	Elateridae	100
	Byrrhidae	50
	Helodidae	50
	Cerambycidae	100
	Cerambycidae, _Toxotus_ ♂	150
	Cerambycidae, _Toxotus_ ♀	250
	Chrysomelidae	55
	Curculionidae	25
	Scarabeidae	100
Hymenoptera	(Symphyta) (unid.)	100
	Tenthredinidae	100
	(Apocrita) (unid.)	50
	Braconidae	50
	Ichneumonidae	25
	Formicidae	15
	Vespidae	50

258

APPENDIXES

Appendix C. (cont.)

Order	Family	Calories
Araneae		100
Hydracarina		15
Phalangida		50
Solpugida		50
Isopoda		50
Chilopoda		100
Pulmonata		150
Annelida		50
Nematoda		25

APPENDIXES

1. Percentage of prey delivered to nestling Yellowheads $\geq$ 50 calories, Turnbull National Wildlife Refuge. Values are higher on more productive lakes and during the emergence period.

Lake	Dawn-0700	0700-1300	1300-dusk	Weighted average
Lower Turnbull	35.1	84.2	42.3	57.6
Isaacson	66.7	86.7	82.7	80.9
Blackhorse	23.7	76.4	74.1	64.9
Big McDowell	-	90.4	84.5	(87.5)
Kepple	76.4	99.7	94.1	92.8
Thirty Acre	89.9	71.8	48.5	66.1

2. Percentage of samples taken from nestling Yellowheads with only prey $\geq$ 50 calories. Turnbull National Wildlife Refuge.

Lake	Dawn-0700	0700-1300	1300-dusk	Weighted average	Overall lake productivity
Lower Turnbull	27.3	39.7	24.6	31.2	Usually high
Isaacson	18.2	68.4	56.5	53.6	Moderately high
Blackhorse	25.0	60.0	40.0	45.0	Intermediate
Big McDowell	-	56.3	33.3	(44.8)	Moderately high
Kepple	50.0	95.8	68.2	75.6	Very high
Thirty Acre	25.0	45.5	27.3	34.1	Moderately high

Appendix E. Size distribution of prey delivered to nestling Yellowheads, Turnbull National Wildlife Refuge, 1968. Captures per hour of prey ≥ 50 calories (in parentheses) are higher in mixed samples than in samples with only large prey.

Time of Day	Only prey ≥ 50 cal		Both large & small prey			Only prey < 50 cal		Difference in rate of delivery of large prey in mixed vs. pure samples
	# samples	# prey	# samples	# prey ≥50 cal	# prey <50 cal	# samples	# prey	
Dawn–0700	1	12(12.0)	1	30(30.6)	8	0	0	+18.6
0700–1300	3	47(15.7)	8	316(39.5)	42	0	0	+23.8
1300–dusk	12	179(14.9)	12	220(18.3)	18	0	0	+3.4
Total	16	238	21	566	68	0	0	
Average delivery rate of large prey		(14.9/hr)		(27.0/hr)				

Appendix F. Size distribution of prey delivered to nestling Redwings, Turnbull National Wildlife Refuge. Captures per hour of prey $\geq$ 50 calories (in parentheses) are higher in mixed samples than in samples with only large prey in 2 out of 3 time periods.

Time of Day	Only prey $\geq$ 50 cal		Both large & small prey			Only prey < 50 cal		Difference in rate of delivery of large prey in mixed vs. pure samples
	# samples	# prey	# samples	# prey $\geq$50 cal	# prey <50 cal	# samples	# prey	
Dawn–0700	2	5(2.5/hr)	7	39(5.6/hr)	40	1	7	+3.1
0700–1300	48	723(15.1/hr)	54	636(11.8/hr)	334	7	49	−3.3
1300–dusk	26	153(5.9/hr)	85	594(7.0/hr)	761	7	20	+1.1
Total	76	881	146	1269	1135	15	76	
Average delivery rate of large prey		(11.6/hr)		(8.7/hr)				

APPENDIXES

Appendix G.

1. Percentage of prey delivered to nestling blackbirds ≥ 50 calories, Columbia National Wildlife Refuge.

Species		Dawn–0700	0700–1300	1300–dusk	Weighted average
Redwing	1964–65	36.4	82.1	40.8	56.4
Redwing	1968	–	62.8	66.4	(64.6)
Yellowhead	1963	81.6	99.3	91.6	92.8
Yellowhead	1964–65	50.8	80.7	71.4	71.0
Yellowhead	1968	27.7	92.7	87.0	77.4

2. Percentage of samples with only prey ≥ 50 calories, Columbia National Wildlife Refuge.

Species		Dawn–0700	0700–1300	1300–dusk	Weighted average
Redwing	1964–65	22.2	53.0	27.3	36.6
Redwing	1968	–	41.7	40.0	(40.9)
Yellowhead (Lyle Lake only)	1963	66.7	81.3	72.0	74.7
Yellowhead	1964–65	10.0	43.5	39.1	35.0
Yellowhead	1968	12.5	50.0	40.0	38.5

263

Appendix H. Size distribution of prey delivered to nestling Yellowheads, Columbia National Wildlife Refuge. Numbers of prey ≥ 50 calories/hour sampling are given in parentheses.

1. 1963

	Only prey ≥ 50 cal # samples	# prey	Both large & small prey # samples	# prey ≥50 cal	# prey <50 cal	Only prey < 50 cal # samples	# prey	Difference in rate of delivery of large prey in mixed vs. pure samples
Dawn-0700	18	159(8.8)	9	81(9.0)	54	1	3	+0.2
0700-1300	13	415(31.9)	3	121(40.3)	4	0	0	+8.4
1300-dusk	18	224(12.4)	7	116(16.6)	31	0	0	+4.2
Total	49	798	19	318	89	1	3	
Average delivery rate of large prey		(16.3/hr)		(16.7/hr)				

2. 1964-65

Time of Day	Only prey ≥ 50 cal # samples	# prey	Both large & small prey # samples	# prey ≥50 cal	# prey <50 cal	Only prey < 50 cal # samples	# prey	Difference in rate of delivery of large prey in mixed vs. pure samples
Dawn-0700	1	2(2.0)	9	118(12.6)	111	1	16	+10.6
0700-1300	10	157(15.7)	13	133(17.9)	93	0	0	+2.2
1300-dusk	9	159(17.7)	14	275(19.0)	242	0	0	+1.3
Total	20	318	36	626	446	1	16	
Average delivery rate of large prey		(15.9/hr)		(17.4/hr)				

Appendix H. (cont.)

3. 1968

Time of Day	Only prey ≥ 50 cal		Both large & small prey			Only prey < 50 cal		Difference In rate of delivery of large prey in mixed vs. pure samples
	# samples	# prey	# samples	# prey ≥50 cal	# prey <50 cal	# samples	# prey	
Dawn-0700	1	1(1.0)	7	149(2.3)	392	0	0	+20.3
0700-1300	11	289(26.3)	11	281(25.5)	45	0	0	−0.8
1300-dusk	8	106(13.3)	12	504(42.0)	91	0	0	+28.7
Total	20	396	30	934	528	0	0	
Average delivery rate of large prey		(19.8/hr)		(31.1/hr)				

Appendix I. Size distributions of prey delivered to nestling Redwings, Columbia National Wildlife Refuge. Numbers of prey ≥ 50 calories/hour of sampling are given in parentheses.

1. 1964-65

Time of Day	Only prey ≥ 50 cal # samples	Only prey ≥ 50 cal # prey	Both large & small prey # samples	Both large & small prey # prey ≥50 cal	Both large & small prey # prey <50 cal	Only prey < 50 cal # samples	Only prey < 50 cal # prey	Difference in rate of delivery of large prey in mixed vs. pure samples
Dawn-0700	4	15(3.8)	14	158(11.3)	310	0	0	+7.5
0700-1300	18	146(8.1)	16	294(18.4)	96	0	0	+10.3
1300-dusk	9	51(5.7)	24	321(13.4)	526	1	40	+7.7
Total	31	212	54	773	932	1	40	
Average delivery rate of large prey		(6.8/hr)		(14.3/hr)				

2. 1968

Time of Day	Only prey ≥ 50 cal # samples	Only prey ≥ 50 cal # prey	Both large & small prey # samples	Both large & small prey # prey ≥50 cal	Both large & small prey # prey <50 cal	Only prey < 50 cal # samples	Only prey < 50 cal # prey	Difference in rate of delivery of large prey in mixed vs. pure samples
Dawn-0700	–	–	–	–	–	–	–	–
0700-1300	5	47(9.4)	7	66(9.4)	67	1	5	0.0
1300-dusk	6	52(8.7)	9	104(11.6)	79	0	0	+2.9
Total	11	99	16	170	146	1	5	
Average delivery rate of large prey		(9.0/hr)		(10.6/hr)				

APPENDIXES

Appendix J. Sites of nests of Yellow-headed Blackbirds in Washington and British Columbia.

Location	Year	Site			
		Emergent aquatic	Shrubs over water	Shrubs over dry land	Trees
Turnbull Refuge	1964	66	0	0	0
	1965	102	0	0	0
	1966	61	0	0	0
Columbia Refuge	1963	84	0	0	0
	1964	106	0	0	0
	1965	86	0	0	0
	1966	40	0	0	0
	1968	84	0	0	0
Cariboo Parklands	1963	136	0	0	0
	1964	157	0	0	0
Total		922	0	0	0

APPENDIXES

Appendix K.   Sites of nests of Red-winged Blackbirds in Washington, British Columbia and Costa Rica.

Location	Year	Site			
		Emergent aquatic	Shrubs over water	Shrubs over dry land	Trees
Seattle	1963	49	0	0	0
	1965	28	0	0	0
	1968	24	0	0	0
Turnbull Refuge	1964	118	0	0	0
	1965	162	3	0	12
	1966	169	0	0	0
	1967	153	0	0	0
Columbia Refuge	1963	17	0	10	0
	1964	79	0	0	0
	1965	95	0	1	1*
	1968	107	0	4	0
Crab Creek	1963	36	0	0	0
	1964	39	0	0	0
Cariboo Parklands	1963-64	19	0	0	0
Taboga, Costa Rica	1967	90	3	0	0
Total		989	6	15	13

*in an old Northern Oriole nest!

Appendix L. Clutch sizes in Yellow-headed Blackbirds.

Locality	Year	Number Nests	Clutch Size					X̄	s²	Source
			1	2	3	4	5			
Columbia National Wildlife Refuge, Washington	1964	63		2	17	41	3	3.71	.369	Orians, this study
	1965	37		1	4	30	2	3.89	.266	"
	1968	55	1		16	37	1	3.67	.372	"
Turnbull National Wildlife Refuge, Washington	1962	134		10	27	94	3	3.67	.418	Willson, 1966
	1963	132			12	112	8	3.97	.152	"
	1964	69			15	52	2	3.81	.214	Orians, this study
	1965	18			4	14		3.78	.183	"
	1968	19		3	8	8		3.26	.538	"
Rush Lake, B.C.	1964	44		1	20	23		3.50	.302	"
Westwick Lake, B.C.	1964	18		3	8	7		3.22	.536	"
Saskatoon, Sask.	1966	85						3.6		Miller, 1968
Utah	1941	118						3.7		Fautin, 1941
Minneapolis, Minn.	1909	26			2	22	2	4.0	.160	Roberts, 1909
NW Iowa	1938	504						3.1		Ammann, 1938
Stoddard, Wis.	1959	44			6	38		3.86	.121	Young, 1963
	1960	26			5	21		3.81	.162	"
	1963	98						3.30		"

Appendix M. Clutch sizes in Red-winged Blackbirds.

Locality	Year	Number Nests	1	2	3	4	5	6	X̄	S²	Source
Seattle, Washington	1963	23			18	5			3.22	.178	Orians
"	1965	24		2	15	7			3.21	.346	"
"	1968	13			8	4	1		3.46	.436	"
Crab Creek, Wn.	1963	21			4	14	3		3.95	.348	"
"	1964	29			5	19	5		4.00	.357	"
Columbia National Wildlife Refuge	1963	13				10	3		4.23	.192	"
Washington	1964	58		1	7	34	16		4.12	.459	"
	1965	34				24	10		4.29	.214	"
	1968	52			18	34			3.64	.393	"
Turnbull National Wildlife Refuge	1961	12		1	4	6	1		3.58	.629	"
Washington	1962	52		4	15	30	3		3.62	.516	"
	1964	51		2	13	31	5		3.76	.624	"
	1965	124			27	73	24		3.98	.414	"
	1966	155		9	53	82	9	2	3.63	.547	"
	1967	125		5	22	86	12		3.84	.410	"
	1968	34		1	14	17	2		3.59	.431	"
Haskell Ranch, Calif.	1960	20		1	9	10			3.45	.366	"
E. Park Reservoir, Cal.	1960	85		2	23	55	5		3.75	.403	"
Belvedere, Alberta	?	30	1			21	7	2	4.37	.378	Henderson fide Bent, 1958
Saskatoon, Sask.	1966	46		1	7	27	11		4.02	.600	Miller (pers. comm.)
"	1967	27			5	15	6	1	4.11	.641	"
"	1968	93	6	3	21	50	11	2	3.68	1.05	"
Wisconsin	1950	153							3.7		Beer & Tibbits, 1950
" (Stoddard)	1959	135		4	29	100	2		3.74	.283	Young, 1963
"	1960	191		3	41	136	11		3.81	.301	"
"	1963	432							3.17		"
Ithaca, New York	1960	450							3.45		Case & Hewitt, 1963
"	1961								3.61		"
Harrisburg, Pa.	1938	27		2	11	14			3.74	2.507	Wood, 1938
Chesapeake Bay	1963	537		24	329	175	9		3.31	.339	Meanley & Webb, 1963
New Jersey, tidal marsh	1975	164	10	22	86	45	1		3.03	.680	Caccamise, 1976
N. central Oklahoma	1967	243							3.4		Goddard & Board, 1967
Stuttgart, Ark.	1957	100							3.2		Neff & Meanley, 1957
Hacienda Taboga,	1966	46		4	40	2			2.96	.131	Orians, 1973
Costa Rica	1967	104	5	25	73	1			2.67	.339	"

APPENDIXES

Appendix N. Food delivered to nestling blackbirds at Pinamar, Argentina, October-November, 1973. Order totals are given in parentheses.

Prey	Pseudoleistes virescens		Agelaius thilius		Amblyramphus holosericeus		Total	
	#	%	#	%	#	%	#	%
Odonata							(55)	(13.3)
Anisoptera			17	15.2	5	5.1	22	5.3
Zygoptera			38	33.9			38	9.2
Orthoptera							(23)	(5.6)
Tettigoniidae	2	1.0			10	10.1	12	2.9
Gryllidae	11	5.4					11	2.7
Dermaptera					1	1.0	1	0.2
Hemiptera							(7)	(1.6)
Belostomatidae			4	3.6			4	1.0
Pentatomidae	1	0.5			1	1.0	2	0.4
? Adult			1	0.9			1	0.2
Homoptera							(2)	(0.4)
Nymphs	1	0.5					1	0.2
Adults					1	1.0	1	0.2
Diptera							(12)	(2.9)
Syrphidae (A)*			1	0.9	1	1.0	2	0.4
? Larvae			8	7.1			8	1.9
? Pupae			2	1.8			2	0.4
Coleoptera							(38)	(9.1)
Carabidae (A)	2	1.0	3	2.7			5	1.2
Dytiscidae (L)*	1	0.5	1	0.9			2	0.4
Helodidae (L)			2	1.8	16	16.2	18	4.3
Helodidae (A)					1	1.0	1	0.2
Tenebrionidae (A)					3	3.0	3	0.7
Scarabeidae (A)	4	2.0					4	1.0
? Adults	3	1.5	1	0.9	1	1.0	5	1.2
Trichoptera			3	2.7			3	0.7
Lepidoptera							(147)	(34.9)
? Larvae	2	1.0	2	1.8	25	25.3	29	7.0
? Pupae			3	2.7	12	12.1	15	3.6
? Adults	7	3.4	1	0.9			8	1.9
Sp. A Larvae	78	38.2					78	18.8
Sp. B Larvae	7	3.4					7	1.7
Sp. C Larvae	9	4.4					9	2.2
Sp. D Larvae			1	0.9			1	0.2

Appendix N.   (Continued)
Food delivered to nestling blackbirds at Pinamar, Argentina,
October-November, 1973. Order total are given in parentheses.

Prey	Pseudoleistes virescens		Agelaius thilius		Amblyramphus holosericeus		Total	
	#	%	#	%	#	%	#	%
Insecta (Larvae)	2	1.0	1	0.9	6	6.1	9	2.2
Araneida							(91)	(21.9)
Adults	43	21.1	13	11.6	12	12.1	68	16.4
Egg Cases	18	8.8	3	2.7	2	2.0	23	5.5
Gastropoda	1	0.5	3+	2.7			4	1.0
Osteichthyes			3	2.7			3	0.7
Amphibia								
Hylidae	1	0.5			2	2.0	3	0.7
Reptilia								
Anguidae	1	0.5					1	0.2
Scincidae	1	0.5					1	0.2
Totals	195		111		99		405	
H		.844		.997		.778		1.190
H_{max}		1.300		1.323		1.216		1.548

*A = adults; L = larvae

APPENDIXES

Appendix O.   Scientific names of Argentine birds mentioned in Table 8.2.

Family	Common Name	Scientific Name
Furnariidae	Curve-billed Reedhaunter	*Limnornis curvirostris*
	Wren-like Rushbird	*Phleocryptes melanops*
	Sulphur-bearded Spinetail	*Cranioleuca sulphurifera*
	Bay-capped Wren Spinetail	*Spartanoica maluroides*
Tyrannidae	Black-and-white Monjita	*Xolmis dominicana*
	Spectacled Tyrant	*Hymenops perspicillata*
	Warbling Doradito	*Pseudocolopteryx flaviventris*
	Many-colored Rush-tyrant	*Tachuris rubrigastra*
Motacillidae	Correndera Pipit	*Anthus correndera*
Icteridae	Yellow-winged Blackbird	*Agelaius thilius*
	Scarlet-headed Blackbird	*Amblyramphus holosericeus*
	Brown-and-yellow Marshbird	*Pseudoleistes virescens*
Fringillidae	Grassland Yellow-finch	*Sicalis luteola*
	Rufous-collared Sparrow	*Zonotrichia capensis*
	Long-tailed Reed-finch	*Donacospiza albifrons*
	Great Pampa-finch	*Embernagra platensis*

273

APPENDIXES

Appendix P. Scientific names of North American birds mentioned in Table 8.2.

Family	Common Name	Scientific Name
Troglodytidae	Long-billed Marsh Wren	_Telmatodytes palustris_
	Short-billed Marsh Wren	_Cistothorus platensis_
Parulidae	Yellowthroat	_Geothlypis trichas_
Icteridae	Brown-headed Cowbird	_Molothrus ater_
	Red-winged Blackbird	_Agelaius phoeniceus_
	Yellow-headed Blackbird	_Xanthocephalus xanthocephalus_
	Tricolored Blackbird	_Agelaius tricolor_
	Western Meadowlark	_Sturnella neglecta_
	Bobolink	_Dolichonyx oryzivorus_
	Common Grackle	_Quiscalus quiscula_
	Brewer's Blackbird	_Euphagus cyanocephalus_
Fringillidae	Savannah Sparrow	_Passerculus sandwichensis_
	Clay-colored Sparrow	_Spizella pallida_
	Swamp Sparrow	_Melospiza georgiana_
	Song Sparrow	_Melospiza melodia_

References

Alchian, A. A., and W. R. Allen. 1964. *University Economics*. Wadsworth Publishing Co., Belmont, Calif.

Allen, A. A. 1914. The Red-winged Blackbird: A study of the ecology of a cat-tail marsh. Abstr. *Proc. Linn. Soc. New York* Nos. 24-25 (1911-1913):43-128.

Altmann, S. A., S. S. Wagner and S. Lenington. 1977. Two models for the evolution of polygyny. *Behav. Ecol. & Sociobiol.* 2:397-410.

Ammann, G. A. 1938. The life history and distribution of the Yellow-headed Blackbird. Ph.D. Thesis, Univ. of Michigan.

Balda, R. P. 1969. Foliage use by birds of the oak-juniper woodland and ponderosa pine forest in southeastern Arizona. *Condor* 71:399-412.

Bannerman, D. A. 1953. *The Birds of West and Equatorial Africa*. Vol. II. Oliver and Boyd, Edinburgh and London.

Beecher, W. J. 1942. Nesting birds and the vegetation substrate. Chicago Ornithological Soc., Chicago Museum Nat. Hist., 1-69.

Beecher, W. J. 1951. Adaptations for food getting in the American blackbirds. *Auk* 68:411-440.

Beer, J. R., and D. Tibbits. 1950. Nesting behavior of the Red-winged Blackbird. *Flicker* 22:61-77.

Bell, B. D., and R. J. Hornby. 1969. Polygamy and nest sharing in the Reed Bunting. *Ibis* 111:402-405.

Bent, A. C. 1958. Life histories of North American blackbirds, orioles, tanagers, and allies. *U.S. Nat. Mus. Bull.* 211.

Bick, G. H., and J. C. Bick. 1965. Demography and behavior of the damselfly, *Argia apicalis* (Say) (Odonata: Coenagrionidae). *Ecology* 46:461-472.

Boucher, D. H. 1977. On wasting parental investment. *Amer. Natur.* 111:786-788.

Bretz, J. H. 1959. Washington's channeled scablands. State of Washington, Division of Mines and Geology. *Bull.* 45:1-57.

Brown, J. L. 1964. The evolution of diversity in avian territorial systems. *Wilson Bull.* 76:160-169.

Brown, J. L. 1966. Types of group selection. *Nature* 211:870.

Brown, J. L., and G. H. Orians. 1970. Spacing patterns in mobile animals. *Ann. Rev. Ecol. & Syst.* 1:239-262.

Burckhardt, D. 1948. Zur Brutbiologie der Beutelmeise, *Remiz pendulinus* (L.) *Orn. Beob.* 45:7-31.

Byers, C. F. 1940. Notes on the emergence and life history of the dragonfly *Pantala flavescens. Proc. Fla. Acad. Sci.* 5:14-25.

Cabrera, A. J. 1968. *Flora de la Provincia de Buenos Aires.* Instituto Nacional de Technología y Agricultura. Buenos Aires.

Caccamise, D. F. 1976. Nesting mortality in the Red-winged Blackbird. *Auk* 93:517-534.

Cade, T. J. 1960. Ecology of the Peregrine and Gyrfalcon populations in Alaska. *Univ. Calif. Publ. Zool.* 63:151-290.

Case, N. A., and O. H. Hewitt. 1963. Nesting and productivity of the Red-winged Blackbird in relation to habitat. *The Living Bird*, Second Annual of the Cornell Laboratory of Ornithology, pp. 7-20.

Catchpole, C. K. 1974. Habitat selection and breeding success in the Reed Warbler (*Acrocephalus scirpaceus*). *J. Anim. Ecol.* 43:363-380.

Charnov, E. L. 1973. Optimal foraging: Some theoretical explorations. Ph.D. dissertation, University of Washington, Seattle.

Charnov, E. L. 1976a. Optimal foraging, the marginal value theorem. *Theoret. Pop. Biol.* 9:129-136.

Charnov, E. L. 1976b. Optimal foraging: Attack strategy of a mantid. *Amer. Natur.* 110:141-151.

Charnov, E. L., G. H. Orians and K. Hyatt. 1976. Ecological implications of resource depression. *Amer. Natur.* 110:247-259.

Claassen, P. W. 1921. *Typha* insects: Their ecological relationships. *Cornell Univ. Agric. Exp. Stat., Memoirs* 47:459-531.

Cody, M. L. 1966. The consistency of inter- and intra-specific continental bird species counts. *Amer. Natur.* 100:371-376.

Cody, M. L. 1968. On the methods of resource division in grassland bird communities. *Amer. Natur.* 102:107-147.

Cody, M. L. 1969. Convergent characteristics in sympatric species: A possible relation to interspecific competition and aggression. *Condor* 71:223-237.

Cody, M. L. 1971. Ecological aspects of reproduction. In: Farner, D. S., and J. R. King (eds.). *Avian Biology*, Vol. 1. Academic Press, New York.

Cody, M. L. 1974. *Competition and the Structure of Bird Communities*. Monographs in Population Biology No. 7, Princeton Univ. Press, Princeton, N.J.

Cook, P. P., and H. S. Horn. 1968. A sturdy trap for sampling emergent Odonata. *Ann. Entom. Soc. Amer.* 61:1506-1507.

Corbet, P. S. 1952. An adult population study of *Pyrrhosoma nymphula* (Sulzer): (Odonata, Coenagrionidae). *J. Anim. Ecol.* 21:206-222.

Corbet, P. S. 1957. The life-history of the Emperor Dragonfly *Anax imperator* Leach (Odonata: Aeshnidae). *J. Anim. Ecol.* 26:1-69.

Corbet, P. S. 1962. *A Biology of Dragonflies*. H. F. and G. Witherby, London.

Crook, J. H. 1964. The evolution of social organization and visual communication in the weaver birds (Ploceinae). *Behav. Suppl*. 10:1-178.

Dawkins, R. and T. R. Carlisle. 1976. Parental investment, mate desertion and a fallacy. *Nature* 262:131-133.

Devitt, O. E. 1964. An extension of the breeding range of Brewer's Blackbird in Ontario. *Canad. Field Natur*. 78:42-46.

Dolbeer, R. A. 1976. Reproductive rate and temporal spacing of nesting of Red-winged Blackbirds in upland habitat. *Auk* 93:343-355.

Donaldson, L. R., P. R. Olson, S. Olsen and Z. F. Short. 1971. The Fern Lake Studies. Univ. Wash. College of Fisheries, Contribution 352.

Drury, W. H., Jr. 1961. Studies of the breeding biology of the Horned Lark, Water Pipit, Lapland Longspur, and Snow Bunting on Bylot Island, Northwest Territories, Canada. *Bird-Banding* 32:1-46.

Dyrcz, A. 1974. Factors affecting the growth rates of nestling Great Reed Warblers and Reed Warblers at Milicz, Poland. *Ibis* 116:330-339.

Edmondson, W. T. 1963. Pacific Coast and Great Basin. In: D. G. Frey (ed.). *Limnology in North America*, pp. 371-392. Univ. Wisconsin Press, Madison, Wisconsin.

Eisner, T. 1970. Chemical defenses against predation in arthropods. In: E. Sondheimer and J. B. Simeone (eds.). *Chemical Ecology*. Academic Press, New York.

Emlen, J. M. 1966. The role of time and energy in food preference. *Amer. Natur*. 100:611-617.

Errington, P. L. 1945. Some contributions of a fifteen year local study of the northern bobwhite to a knowledge of population phenomena. *Ecol. Monogr*. 15:1-34.

Fautin, R. W. 1941. Incubation studies of the Yellow-headed Blackbird. *Wilson Bull.* 53:107-122.

Fretwell, S. D. 1972. *Populations in a Seasonal Environment.* Monographs in Population Biology No. 5, Princeton Univ. Press, Princeton, N.J.

Fretwell, S. D., D. E. Bowen and H. A. Hespenheide. 1974. Growth rates of young passerines and the flexibility of clutch size. *Ecology* 55:907-909.

Fretwell, S. D., and H. L. Lucas. 1969. On territorial behavior and other factors influencing habitat distribution in birds. I. Theoretical development. *Acta Biotheoretica* 19:16-36.

Fry, C. H. 1972. The social organisation of Bee-eaters (Meropidae) and co-operative breeding in hot-climate birds. *Ibis* 114:1-14.

Furrer, R. K. 1974. Nest site stereotypy and optimal breeding strategy in a population of Brewer's Blackbirds (*Euphagus cyanocephalus*). Aku-Fotodruck, Zurich.

Gambles, R. M. 1963. The larval stages of Nigerian dragonflies, their biology and development. *J. West African Sci. Assoc.* 8:111-120.

Goddard, S. V., and V. V. Board. 1967. Reproductive success of Red-winged Blackbirds in north central Oklahoma. *Wilson Bull.* 79:283-289.

Grant, P. R. 1965. A systematic study of the terrestrial birds of the Tres Marías Islands, Mexico. *Postilla, Peabody Mus. Nat. Hist.* No. 90.

Grant, P. R. 1966. Further information on the relative length of the tarsus in land birds. *Postilla, Peabody Mus. Nat. Hist.* No. 98.

Grant, P. R. 1968. Polyhedral territories of animals. *Amer. Natur.* 102:75-80.

Grieve, E. G. 1937. Studies on the biology of the damselfly, *Ischnura verticalis* Say, with notes on certain parasites. *Ent. Amer.* 17:121-153.

Hamilton, T. H. 1961. On the functions and causes of sexual dimorphism in the breeding plumage characters of North American species of warblers and orioles. *Amer. Natur.* 95:121-123.

Hamilton, W. D. 1964. The genetical evolution of social behavior. *J. Theoret. Biol.* 7:1-52.

Harris, R. D. 1944. The Chestnut-collared Longspur in Manitoba. *Wilson Bull.* 56:105-115.

Hindwood, K. A., and A. R. McGill. 1958. *The Birds of Sydney.* Royal Zool. Soc. New South Wales.

Holcomb, L. C. 1974. The question of possible surplus females in Red-winged Blackbirds. *Wilson Bull.* 86:177-179.

Holcomb, L. C., and G. Twiest. 1968. Ecological factors affecting nest building in Red-winged Blackbirds. *Bird-Banding* 39:14-22.

Holdridge, L. R. 1957. *Life Zone Ecology.* Tropical Science Center, San José, Costa Rica.

Holling, C. S. 1959. The components of predation as revealed by a study of small mammal predation of the European Pine Sawfly. *Canad. Ent.* 91:293-320.

Holling, C. S. 1965. The functional response of predators to prey density and its role in mimicry and population regulation. *Ent. Soc. Canad. Mem.* No. 45.

Holling, C. S. 1966. The functional response of invertebrate predators to prey density. *Ent. Soc. Canad. Mem.* No. 48.

Holm, C. H. 1973. Breeding sex ratios, territoriality, and reproductive success in the Red-winged Blackbird (*Agelaius phoeniceus*). *Ecology* 54:356-365.

Horn, H. S. 1968. The adaptive significance of colonial nesting in the Brewer's Blackbird (*Euphagus cyanocephalus*). *Ecology* 49:682-694.

Horn, H. S. 1970. Social behavior of nesting Brewer's Blackbirds. *Condor* 72:15-23.

Horn, H. S. 1971. *The Adaptive Geometry of Trees.* Monographs in Population Biology No. 3, Princeton Univ. Press, Princeton, N.J.

Hudson, W. H. 1923. *Birds of La Plata*. J. M. Dent & Sons, London.

Kahl, M. P. 1964. Food ecology of the Wood Stork (*Mycteria americana*) in Florida. *Ecol. Monogr.* 34:97-117.

Karr, J. R. 1971. Structure of avian communities in selected Panama and Illinois habitats. *Ecol. Monogr.* 41:207-233.

King, J. R. 1972. Adaptive periodic fat storage of birds. *Proc. XV Intern. Ornith. Congr.* pp. 200-217.

Kormondy, E. J., and J. L. Gower. 1965. Life history variations in an association of Odonata. *Ecology* 46:882-886.

Krebs, J. R., J. Ryan and E. L. Charnov. 1974. Hunting by expectation or optimal foraging? A study of patch use by chickadees. *Anim. Behav.* 22:953-964.

Kushlan, J. A. 1976. Wading bird predation in a seasonally fluctuating pond. *Auk* 93:464-476.

Lack, D. 1948. Natural selection and family size in the Starling. *Evolution* 2:95-110.

Lack, D. 1954a. *The Natural Regulation of Animal Numbers*. Oxford Univ. Press, Oxford.

Lack, D. 1954b. The evolution of reproductive rates. In: Huxley, J., A. C. Hardy and E. B. Ford (eds.). *Evolution as a Process*. Allen and Unwin, London.

Lack, D. 1965. *Enjoying Ornithology*. Methuen, London.

Lack, D. 1966. *Population Studies of Birds*. Clarendon Press, Oxford.

Lack, D. 1968. *Ecological Adaptations for Breeding in Birds*. Methuen, London.

Levins, R. 1968. *Evolution in Changing Environments*. Monographs in Population Biology No. 2, Princeton Univ. Press, Princeton, N.J.

Lind, E. M., and M. E. S. Morrison. 1974. *East African Vegetation*. Longmans, London.

REFERENCES

Lorenz, K. Z. 1949. Über die Beziehungen zwischen Kopfform und Zirkelbewegung bei Sturniden und Ikteriden. In: Mayr, E. (ed.). *Ornithologie als biologische Wissenschaft*. Heidelberg, Carl Winter.

Lutz, P. 1968. Life history studies on *Lestes eurinus* Say (Odonata). *Ecology* 49:576-579.

Lutz, P. E., and A. R. Pittman. 1970. Some ecological factors influencing a community of adult Odonata. *Ecology* 51:279-284.

Macan, T. T. 1964. The Odonata of a moorland fishpond. *Int. Revue ges. Hydrobiol.* 49:325-360.

Macan, T. T. 1966a. Predation by *Salmo trutta* in a moorland fishpond. *Verh. Internat. Verein. Limnol.* 16:1081-1087.

Macan, T. T. 1966b. The influence of predation on the fauna of a moorland fishpond. *Arch. Hydrobiol.* 61:432-452.

MacArthur, R. H. 1969. Species packing and what interspecies competition minimizes. *Proc. Nat. Acad. Sci.* 64:1369-1371.

MacArthur, R. H. 1972. *Geographical Ecology*. Harper & Row, New York.

MacArthur, R. H., and J. W. MacArthur. 1961. On bird species diversity. *Ecology* 42:594-598.

MacArthur, R. H., J. W. MacArthur and J. Preer, 1962. On bird species diversity. II. Prediction of bird census from habitat measurements. *Amer. Natur.* 96:167-174.

MacArthur, R. H., and E. R. Pianka. 1966. On optimal use of a patchy environment. *Amer. Natur.* 100:603-609.

MacArthur, R. H., H. Recher and M. L. Cody. 1966. On the relation between habitat selection and species diversity. *Amer. Natur.* 100:319-332.

McKee, B. 1972. *Cascadia*. The Geological Evolution of the Pacific Northwest. McGraw-Hill, New York.

Main, B. Y. 1957. Biology of aganippine trapdoor spiders (Mygalomorphae: Ctenizidae). *Austral. J. Zool.* 5:402-473.

Maynard Smith, J. 1958. *The Theory of Evolution.* Penguin Books, Harmondsworth, Middlesex.

Maynard Smith, J. 1964. Group selection and kin selection. *Nature* 201:1145-1147.

Maynard Smith, J. 1974. *Models in Ecology.* Cambridge Univ. Press, Cambridge.

Meanley, B., and J. S. Webb. 1963. Nesting ecology and reproductive rate of the Red-winged Blackbird in tidal marshes of the Upper Chesapeake Bay region. *Chesapeake Science* 4:90-100.

Medawar, P. B. 1957. *The Uniqueness of the Individual.* Methuen, London.

Miller, R. S. 1968. Conditions of competition between Redwings and Yellowheaded Blackbirds. *J. Anim. Ecol.* 37:43-62.

Mock, D. W. 1975. Feeding methods of the Boat-billed Heron, a deductive hypothesis. *Auk* 92:590-592.

Moreau, R. E. 1966. *The Bird Faunas of Africa and its Islands.* Academic Press, London and New York.

Neff, J. A. 1937. Nesting distribution of the Tri-colored Red-wing. *Condor* 39:61-81.

Neff, J. A., and B. Meanley. 1957. Blackbirds and the Arkansas rice crop. *Arkansas Agric. Exp. Stat. Bull.* 584.

Nero, R. W. 1956a. A behavior study of the Red-winged Blackbird. *Wilson Bull.* 68:5-37, 129-150.

Nero, R. W. 1956b. Red-wing nesting in bird house. *Auk* 73:284.

Nice, M. M. 1943. Studies in the life history of the Song Sparrow. II. *Trans. Linn. Soc. N.Y.* 6:1-328.

Orians, G. H. 1961. The ecology of blackbird (*Agelaius*) social systems. *Ecol. Monogr.* 31:285-312.

Orians, G. H. 1962. Natural selection and ecological theory. *Amer. Natur.* 96:257-263.

Orians, G. H. 1966. Food of nestling Yellow-headed Blackbirds, Cariboo Parklands, British Columbia. *Condor* 68:321-337.

Orians, G. H. 1969. On the evolution of mating systems in birds and mammals. *Amer. Natur.* 103:589-603.

Orians, G. H. 1972. The adaptive significance of mating systems in the Icteridae. *Proc. XV Intern. Ornith. Congr.* pp. 389-398.

Orians, G. H. 1973. The Red-winged Blackbird in tropical marshes. *Condor* 75:28-42.

Orians, G. H., and G. M. Christman. 1968. A comparative study of the behavior of Red-winged, Tricolored, and Yellow-headed Blackbirds. *Univ. Calif. Publ. Zool.* 84:1-81.

Orians, G. H., and G. Collier. 1963. Competition and blackbird social systems. *Evolution* 17:449-459.

Orians, G. H., and H. S. Horn. 1969. Overlap in foods and foraging of four species of blackbirds in the Potholes of central Washington. *Ecology* 50:930-938.

Orians, G. H., C. E. Orians and K. J. Orians. 1977. Helpers at the nest in some Argentine blackbirds. In: B. Stonehouse and C. Perrins (eds.). *Evolutionary Ecology*, Macmillan, London, pp. 137-151.

Orians, G. H., and N. E. Pearson. 1979. On the theory of central place foraging. In: *Analysis of Ecological Systems*, D. J. Horn, R. D. Mitchell and G. R. Stairs (eds.). Ohio State Univ. Press, Columbus, pp. 155-177.

Orians, G. H., and M. F. Willson. 1964. Interspecific territories of birds. *Ecology* 45:736-745.

Otte, D. 1975. On the role of intraspecific deception. *Amer. Natur.* 109:239-242.

Pajunen, V. I. 1962. Studies on the population ecology of

Leucorrhinia dubia V. D. Lind (Odonata: Libellulidae). *Ann. Zool. Soc. 'Vanamo'* 24:1-79.

Payne, R. B. 1969. Breeding seasons and reproductive physiology of Tricolored Blackbirds and Red-winged Blackbirds. *Univ. Calif. Publ. Zool.* 90:1-137.

Payne, R. B. 1974. The evolution of clutch size and reproductive rates in parasitic cuckoos. *Evolution* 28:169-181.

Pearson, N. E. 1974. Optimal foraging theory. Quant. Sci Paper No. 39, Center for Quant. Sci. in Forestry, Fisheries, and Wildlife. Univ. Wash., Seattle.

Perrins, C. M. 1965. Population fluctuations and clutch size in the Great Tit, *Parus major* L. *J. Anim. Ecol.* 34:601-647.

Perrins, C. M. 1970. The timing of birds' breeding seasons. *Ibis* 112:242-255.

Pianka, E. R. 1973. The structure of lizard communities. *Ann. Rev. Ecol. & Syst.* 4:53-74.

Pitelka, F. A. 1958. Timing of molt in Stellar Jays of the Queen Charlotte Islands, British Columbia. *Condor* 60:38-49.

Pitelka, F. A., P. Q. Tomich, and G. W. Treichel. 1955. Ecological relations of jaegers and owls as lemming predators near Barrow, Alaska. *Ecol. Monogr.* 25:85-117.

Pulliam, H. R. 1974. On the theory of optimal diets. *Amer. Natur.* 108:59-74.

Rapport, D. J. 1971. An optimization model of food selection. *Amer. Natur.* 105:575-587.

Rawson, D. S. 1961. A critical analysis of the limnological variables used in assessing the productivity of northern Saskatchewan lakes. *Verh. int. Verein. Limnol.* 14:160-166.

Ridgway, R. 1902. *The Birds of Middle and North America.*

Vol. II. *U.S. Nat. Mus. Bull.* 50. U.S. Govt. Printing Office, Washington, D.C.

Robert, P. A. 1958. *Les Libellules (Odonates)*. Delachoux & Niestle, Neuchatel.

Roberts, T. S. 1909. A study of a breeding colony of Yellow-headed Blackbirds. *Auk* 24:371-389.

Robertson, R. J. 1972. Optimal niche space of the Red-winged Blackbird (*Agelaius phoeniceus*). I. Nesting success in marsh and upland habitat. *Canad. J. Zool.* 50:247-263.

Robertson, R. J. 1973a. Optimal niche space of the Red-winged Blackbird: Spatial and temporal patterns of nesting activity and success. *Ecology* 54:1085-1093.

Robertson, R. J. 1973b. Optimal niche space of the Red-winged Blackbird. III. Growth rate and food of nestlings in marsh and upland habitat. *Wilson Bull.* 85:209-222.

Rodriguez, E. and D. A. Levin. 1976. Biochemical parallels of repellents and attractants in higher plants and arthropods. In: J. W. Wallace and R. L. Mansell (eds.). *Recent Advances in Phytochemistry*, Vol. 10. Biochemical Interaction between Plants and Insects. Plenum Press, New York.

Root, R. B. 1966. The avian response to a population outbreak of the tent caterpillar, *Malacosoma constrictum* (Stretch) (Lepidoptera: Lasiocampidae). *Pan-Pacific Entomologist* 42:48-53.

Rosenzweig, M. L. 1968. Net primary productivity of terrestrial communities: Prediction from climatological data. *Amer. Natur.* 102:67-74.

Rothstein, S. I. 1975. An experimental and teleonomic investigation of avian brood parasitism. *Condor* 77:250-271.

Ruwet, J. C. 1965. *Les Oiseaux des Plaines et du Lac-Barrage de la Lufira Supérieure (Katanga Méridional)*. Fondation de L'Université de Liège pour les recherches scientifiques en Africa Centrale. Université de Liège.

Samuelson, P. A. 1970. *Economics*. 8th edition. McGraw-Hill, New York.

Sawyer, M., and M. I. Dyer. 1968. Yellow-headed Blackbird nesting in southern Ontario. *Wilson Bull*. 80:236-237.

Schoener, T. W. 1970. Size patterns in West Indian *Anolis* lizards. II. Co-relations with sizes of particular sympatric species—displacement and convergence. *Amer. Natur*. 104:155-174.

Schoener, T. W. 1971. Theory of feeding strategies. *Ann. Rev. Ecol. & Syst*. 2:369-404.

Schoener, T. W., and D. H. Janzen. 1968. Notes on environmental determinants of tropical versus temperate insect size patterns. *Amer. Natur*. 102:207-224.

Searcy, W. A. 1977. The effect of sexual selection on male Red-winged Blackbirds (*Agelaius phoeniceus*). Ph.D. Thesis, Univ. of Washington.

Selander, R. K. 1966. Sexual dimorphism and differential niche utilization in birds. *Condor* 68:113-151.

Skutch, A. F. 1935. Helpers at the nest. *Auk* 52:257-273.

Smith, H. M. 1943. Size of breeding populations in relation to egg laying and reproductive success in the Eastern Red-wing (*Agelaius p. phoeniceus*). *Ecology* 24:183-207.

Smythies, B. E. 1953. *The Birds of Burma*. Oliver & Boyd, Edinburgh and London.

Snelling, J. C. 1968. Overlap in feeding habits of Red-winged Blackbirds and Common Grackles nesting in a cattail marsh. *Auk* 85:560-585.

Terborgh, J. 1973. On the notion of favorableness in plant ecology. *Amer. Natur*. 107:481-501.

Tinbergen, N., M. Impekoven and D. Franck. 1967. An experiment on spacing-out as a defence against predation. *Behaviour* 28:307-321.

Trivers, R. L. 1972. Parental investment and sexual selection. In: B. Campbell (ed.). *Sexual Selection and the Descent of Man, 1871-1971.* Aldine Press, Chicago.

Verner, J. 1964. Evolution of polygamy in the Long-billed Marsh Wren. *Evolution* 18:252-261.

Verner, J. 1965. Breeding biology of the Long-billed Marsh Wren. *Condor* 67:6-30.

Verner, J., and G. H. Engelson. 1970. Territories, multiple nest building, and polygyny in the Long-billed Marsh Wren. *Auk* 87:557-567.

Verner, J., and M. F. Willson. 1966. The influences of habitats on mating systems of North American passerine birds. *Ecology* 47: 143-147.

Vervoorst, F. B. 1967. *La Vegetación de la República Argentina. VII. Las comunidades vegetales de la Depresión del Salado.* Serie Fitogeográfico No. 7. Instituto Nacional Technología Agropecuaria, Secretaria de Estado de Agricultura y Ganadería de la Nación. Buenos Aires.

Voigts, D. K. 1973. Food overlap of two Iowa marsh icterids. *Condor* 75:392-399.

Walker, E. M. 1953. *The Odonata of Canada and Alaska.* Vol. I. Univ. of Toronto Press, Toronto.

Ward, P. 1965. Feeding ecology of the Black-faced Dioch *Quelea quelea* in Nigeria. *Ibis* 107:173-214.

Weller, M. W. 1967. Notes on some marsh birds of Cape San Antonio, Argentina. *Ibis* 109:391-411.

Weller, M. W., and L. H. Fredrickson. 1974. Avian ecology of a managed glacial marsh. *The Living Bird*, Twelfth Annual of the Cornell Laboratory of Ornithology, pp. 269-291.

Weller, M. W., and C. E. Spatcher. 1965. Role of habitat in the distribution and abundance of marsh birds. Iowa State Univ. Agr. and Home Econ. Exp. Stat., Spec. Report No. 43.

Werner, E. E. 1972. On the breadth of diet in fishes. Ph.D. Thesis, Michigan State University.

Werner, E. E., and D. J. Hall. 1974. Optimal foraging and size selection of prey by the bluegill sunfish (*Lepomis mochrochirus*). *Ecology* 55:1042-1052.

Wetmore, A. 1926. Observations on the birds of Argentina, Paraguay, Uruguay, and Chile. *U.S. Nat. Mus. Bull.* 133.

Wiens, J. A. 1973. Pattern and process in grassland bird communities. *Ecol. Monogr.* 43:237-270.

Wiens, J. A. 1974. Habitat heterogeneity and avian community structure in North American grasslands. *Amer. Midl. Natur.* 91:195-213.

Wilder, G. D., and H. W. Hubbard. 1938. Birds of Northeastern China. Peking Natural History Bulletin.

Williams, G. C. 1966. *Adaptation and Natural Selection*. Princeton Univ. Press, Princeton, N.J.

Willson, M. F. 1966. Breeding ecology of the Yellowheaded Blackbird. *Ecol. Monogr.* 36:51-77.

Wittenberger, J. F. 1976. The ecological factors selecting for polygyny in altricial birds. *Amer. Natur.* 110:779-799.

Wood, H. B. 1938. Nesting of Red-winged Blackbirds. *Wilson Bull.* 50:143-144.

Woolfenden, G. E. 1975. Florida Scrub Jay helpers at the nest. *Auk* 92:1-15.

Young, H. 1963. Age-specific mortality in the eggs and nestlings of blackbirds. *Auk* 80:145-155.

Zahavi, A. 1957. The breeding birds of the Huleh Swamp and Lake (northern Israel). *Ibis* 99:600-607.

Zimmerman, J. L. 1966. Polygyny in the Dickcissel. *Auk* 83:534-546.

Index

Library of Congress Cataloging in Publication Data

Orians, Gordon H
  Some adaptations of marsh-nesting blackbirds.

  (Monographs in population biology ; 14)
  Includes bibliographical references and index.
    1. Blackbirds—Behavior.   2. Marsh ecology.
  3. Adaptation (Biology)   4. Resource partitioning
  (Ecology)   5. Birds—Behavior.   I. Title.
  II. Series.
  QL696.P2475075   598.8'81   79-84005
  ISBN 0-691-08236-7
  ISBN 0-691-08237-5 pbk.